Cork Science and its Applications II

2nd conference on Cork in Science and Applications, May, 22nd & 23rd, 2019, Cork Museum – Museu del Suro, Palafrugell (Girona), Spain

Editors

Patricia Jové Martín and Maria Verdum Virgos

The Catalan Institute of Cork (ICSuro), R+D Department, Palafrugell (Girona), Spain

Peer review statement

All papers published in this volume of "Materials Research Proceedings" have been peer reviewed. The process of peer review was initiated and overseen by the above proceedings editors. All reviews were conducted by expert referees in accordance to Materials Research Forum LLC high standards.

Published as part of the proceedings series
Materials Research Proceedings
Volume 14 (2019)

ISSN 2474-3941 (Print)
ISSN 2474-395X (Online)

ISBN 978-1-64490-040-6 (Print)
ISBN 978-1-64490-041-3 (eBook)

Distributed worldwide by

Materials Research Forum LLC
105 Springdale Lane
Millersville, PA 17551
USA
http://www.mrforum.com

Manufactured in the United State of America
10 9 8 7 6 5 4 3 2 1

Table of Contents

Preface

Welcome speech from Mr. Joan J. Puig, President of Institut Català del Suro:

Good morning to all the presents.

Be all welcome to this hospitable land that is Catalonia.

My first words go to thank the team of the Institut Català del Suro for the enormous effort they have made in the organization of this event.

For me, it is an honor to host this Congress as President of the Institut Català del Suro Foundation.

The Institut Català del Suro has been, since its foundation in 1991 by the Catalan cork companies, the technological center on Cork with greater capacity to spread the benefits of this natural miracle that is cork.

A scientific rigorous and constant work and diffusion of the cork sector, developed in these 28 years by this organization that has helped evolving the entire cork sector, mostly dedicated to the transformation with oenological utilities, but also capable of advancing in new applications of cork.

Today we start a new stage of the Institut Català del Suro as a private Foundation that intends to continue defending cork and its utilities. And in this new phase we are intensively incorporating the primary forest sector to provide forest management, raw material improvement and environmental visibility and sustainability.

For us to start this new and exciting stage was relevant, and we can imagine a better senario that this Scientific Congress we are inaugurating today.

With it we intend to visualize our engagement towards knowledge and innovation.

But we also intend to show again the will to internationalization and the defense of a product and a sector that can and should lead the revolution of the new economy.

Of that economy that we represent: the one that transforms a natural product, the one that socially occupies people in a dignified environment and the one that imagine a sustainable future for our children.

I wish the greatest successes for all attendees, as their work is creating the knowledge for a sustainable future.

I invite to you all to be active ambassadors of the benefits of cork

Good congress to everybody.

Joan J. Puig
President of Institut Català del Suro
May 22, 2019

Committees

Patricia Jové i Martín
Graduated in Biology and Science and Technology of Food from Universitat de Barcelona and Universitat de Girona. Jové is a Doctor in Experimental Sciences and Sustainability, and a specialist in cork focused on the study of the properties and the new applications of cork as a material. Her PhD was focused on the characterisation and evaluation of the capacity of cork and by products on cork industry as biosorbents of pollutants.
Dr. Jové has been involved in more than 5 European projects and 20 national projects and is the author or co-author of 15 publications, always with cork as leit motiv.
Director of R&D&I Department at Institut Català del Suro since 2014, she is in charge of the management and implementation of ICSuro projects. These projects are focused on two fields: Cork-Wine interaction and new applications of cork in design and architecture. The R&D&I department also advise and helps entrepreneurs to develop new cork products.

Maria Verdum Virgos
Graduated in Biology from the University of Girona and Master in Advanced Microbiology from the University of Barcelona.
Ms. Verdum has been working in Catalan Cork Institute (ICSuro) since 2014. I worked as a Laboratory Technician in Cork center Laboratory. In 2016 I had been changed from Quality department (Cork Center Laboratory) to R&D department. Since then, I am a researcher in since 2016. I have been part of 12 Regional, State and European projects. According of aim of ICSuro, all projects are focus on promoting, developing and raising awareness of cork usage. I am co-author of 2 research papers, and 2 more which are in writing. I am author of 3 communications in international scientific conferences and I am co-author of 3 posters. I have been coordination assistant in the conferences of Cork in Science and its Aplications. I had also been involved as a researcher in research group: Genomics and Proteomics of Bacterial Virulence Factors in the Department of Microbiology of the University of Barcelona in 2014 and the Centre de Recerca en Sanitat Animal-Universitat Autonoma de Barcelona (CReSA-UAB) in 2013.

Paulo Lopes
Paulo Lopes is a wine specialist with a strong technical background and solid experience focused on packaging for the wine industry. Graduated in Forestry Engineering from Universidade Técnica de Lisboa (Portugal). In 2005, Paulo received an Enology and Viticulture PhD from the Faculté d'Oenologie de Bordeaux (France). Along with Denis Dubourdieu and Yves Glories, Paulo has made vital contributions in areas such as oxygen permeation of different wine closures and its impact on the chemical and sensory properties of bottled wine. In 2011, Paulo also received a Wine MBA from BEM Management School (Bordeaux, France). Dr. Lopes has done more than 30 publications as an author or co-author in peer-reviewed journals, grape and wine industry publications and proceedings.
Research Manager at Amorim and Irmãos, S.A., since 2006, being responsible for supervision, planning and completion of R&D projects; technical support for Amorim clients and Sales/Marketing team as well as support product development and compliance; application trials and wine educative seminars.

Jose Ramon Gonzalez Adrados
Jose Ramon Gonzalez Adrados received his PhD in Forestry in 1989 from the Technical University of Madrid. Currently he is senior lecturer at the Engineering and Forest and Environmental Management Department, School of School of of Forestry and Natural Resources, Technical University of Madrid (UPM), Spain. He has been teaching subjects related to non-wood forest products since 2000 in both undergraduate and graduate levels. From 2002 to 2012 he has been working at INIA Forest Research Center as senior researcher and head of its Cork Laboratory. He is currently member of the "Wod and Cork Technology" research group, and Faculty member of the PhD program on Engineering and

Management of the Natural Environment, both at UPM. His research interests include characterization, properties and behavior of non-wood based materials, non-wood raw materials production and manufacturing, and cork-wine interaction.

He is author or co-author of several scientific papers and contributions to scientific congress, including 15 listed in Web of Science data bases, and a patent of a device for measuring forces exerted by stoppers in the bottle neck. He is member of cork related standardization committees at national (AENOR CT56/SCT5) and international (ISOTC87) level.

Gabriel Barbeta i Solà

Doctor and Professor in Ecoarchitecture. Director, in five editions, of the Applied BioConstruction Master, and Health and Harmony Habitat Graduate. Professor in IBN Baubiologie Master, UPV Management of projects of Bioconstruction and Restoration of Heritage Masters, and in the different editions Masters of Sustainability, Structures, and Postgraduate UPC courses of energy certification; Founding member AUS, Professional Association for Sustainable Architecture, COAC Group; and active researcher low tech technologies in CATS, Construction Advanced Technologies and Sustainability in Architecture and Construction Engineering Department, UdG. Girona University. Member editor Committee AEN / CTN 41 / SC 10 "Building with raw land" and National Commission CTE- Earth (Construction Code). Founding member ONGD "Social Architecture" and " Ecoarchitecture Net". Author of multiple earth projects, selected in the exhibition of environmental architecture of the CSAE and in the international competition of Tenerife; Endesa, Ecoviure and BAM 2010 awards by the Font del Rieral school; prize 2012 Ecoviure with half-buried housing of CEB; Author of multiple articles and research.

Ricardo Alves de Sousa

Ricardo Alves de Sousa (São Paulo, 7th October, 1977) is currently Assistant Professor at the Department of Mechanical Engineering and coordinates the Division of Plastic Forming, from the GRIDS-TEMA research group.

He has obtained BSc and MSc degrees in Mechanical Engineering from the University of Porto in 2000 and 2003, respectively, being awarded as the top ten student of the 2000 class in Mechanical Engineering. In 2006, he obtained the PhD degree in Mechanical Engineering from the University of Aveiro, Portugal.

In 2011, received the international scientific ESAFORM (European Association of Material Forming) prize for outstanding contributions in the field of material forming. In 2013 received the Innovation Prize from APCOR, the portuguese association for cork material.

He is author of 3 patents and more than 100 papers, 45 of them in International peer reviewed journals.

Sponsors

Generalitat de Catalunya
**Departament d'Agricultura,
Ramaderia, Pesca i Alimentació**

PLATINUM

ajuntament de
palafrugell

J.V
J. VIGAS, S.A.
1887

Centre de la Propietat
Forestal

Agilent Technologies

GOLD

OLLER

Trefinos

Universitat de Girona
**Centre de Documentació
Europea**

europe
direct
Girona

SILVER

UB

MS manuel serra

QualitySuber

PARTNERS

aecork
Associació d'Empresaris Surers
de Catalunya

CONSORCI FORESTAL
DE CATALUNYA

MUSEU DEL
SURO DE
PALAFRUGELL
MUSEU DE LA CIÈNCIA
I DE LA TÈCNICA DE CATALUNYA

MEDIA PARTNERS

RETECORK

AVC
ASSOCIACIÓ VINÍCOLA CATALANA

Cork Science and its Applications II
Materials Research Proceedings **14** (2019) 1-6

Materials Research Forum LLC
https://doi.org/10.21741/9781644900413-1

Stem Diameter and Height as Traits Linked to Cork Quality (Thickness and Porosity) in Cork Oak (*Quercus suber* L.)

AUGUSTA COSTA[1,2 a,b *] and INÊS BARBOSA[2,c]

[1]Instituto Nacional de Investigação Agrária e Veterinária, I.P., Av. da República, Quinta do Marquês, 2780-159 Oeiras, Portugal

[2]CENSE – Center for Environmental and Sustainability Research, NOVA School of Science and Technology, Campus de Caparica, 2829-516 Caparica, Portugal

[a]augusta.costa@iniav, [b]augusta.costa.sousa@gmail.com, [c]inesalexandrabarbosa@gmail.com

Keywords: Raw Cork Planks, Stem Profile of Cork Quality, Linear Mixed Models

Abstract. Thickness and porosity are properties of raw (natural) cork planks that have strong influence on the end-product quality, particularly on natural cork stoppers or discs, both highly valued natural cork products. In cork oak (*Quercus suber* L.), the between-trees variability of cork quality (thickness and porosity) is relatively well known and has been extensively studied. However, the within-tree variability of these properties has been scarcely addressed in the literature. In this study, we have focused on the within-tree variability of cork quality. Our major working hypothesis is that the within-tree variability of cork quality is related with trees specific growth patterns. We assessed, at tree individual level, the stem height-cork thickness and the stem height-cork porosity relationships that, so far, were not clear. Eventually, the patterns of radial growth, along the stem upwards may have a strong physiological basis. Our results showed: i) a stem height-cork thickness relationship, with trees showing a decrease of the cork thickness with the stem height and; (ii) a stem height-cork porosity relationship, with trees showing a decrease in the cork porosity with the stem height. Forest managers and cork processors require more quantitative information on the inter- and intra-stem variation of cork-thickness and cork-porosity in order to understand the effects of forest management decisions at cork oak woodlands level. This way, the modeled stem profiles of cork quality (thickness and porosity) allow the optimization of high quality raw cork (planks) and determine the economic feasibility of cork harvesting practices.

Introduction

Thickness and porosity are properties of raw (natural) cork planks that have strong influence on the end-product quality, particularly on natural cork stoppers or discs, the most highly valued natural cork products [1].

In the industrial processing of natural cork stoppers, the commercial quality grading classes of cork planks uses primarily the cork thickness as a criterion for differentiating stoppable (commercial cork planks thickness between 27 and 40 mm) and non-stoppable cork planks [2]. Furthermore, several studies showed that the quality profile of natural cork stoppers depends directly on the quality of the raw cork planks [3]. In fact, the quality classes established for natural cork stoppers uses the cork porosity: high-quality cork stoppers have lower porosity (small and/or less lenticels), while poor-quality cork stoppers have higher porosity (large and/or more lenticels) [4].

In cork oak (*Quercus suber* L.), the between-trees variability of cork quality it is relatively well known, and it has been intensively discussed in the literature [5]. However, the within-trees variability of cork thickness and porosity has been scarcely addressed. In previous studies on cork quality variations within the tree, the cork thickness and porosity have been reported to decrease along tree stem upwards, from the base to the crown [6-8].

The overall trends of cork thickness and porosity were considered to be inherent to tree growth, presumably associated with phellogen's intrinsic characteristics, since they occurred in all trees, on all sites. The changes in the cork thickness were associated with differences in the number of cork cells produced rather than with the cells size. Thus, the cork thickness decrease at higher stem heights because the number of suberized cells produced radially by the phellogen in each layer is lower. In contrast, the changes in porosity were associated with differences in the lenticels size rather than in the number of lenticels produced by the lenticular phellogen [9-11].

So far, there is only a general information describing the variation in cork growth and porosity between and within tree, and few, if any, empirical models exist that are able to assess the variation of cork thickness and porosity patterns within individual trees. This may be due to difficulties in having accurate and affordable measurements of cork thickness and porosity on upper-stem diameters which makes the stem height-cork thickness and stem height-cork porosity relationships not extensively studied.

There are good reasons to predict cork properties (thickness and porosity) in the cork oak woodlands and by the forest owners. If cork properties are taken at one first stage, when at the tree, in the forest, it is not too late to correct the stand density, the cork harvesting height or harvesting intensity, or even adjust the cork harvesting period, to improve the cork quality properties. It also may be possible to redirect the cork production to a more valuable and product that is the stoppable good-quality raw cork plank. Nowadays, in the cork flows planning is thus increasingly important to determine the yield of cork but also the yield of cork with predicted properties and correspondent quality.

In this study, we have focused on the within-tree variability of cork quality. Our major working hypothesis is that the within-tree variability of cork quality it is related with trees specific growth patterns. We assessed, at tree individual level, the stem height-cork thickness relationships, presumably associated with phellogen's intrinsic characteristics and; the stem height-cork porosity relationships presumably associated with lenticular phellogen's intrinsic characteristics.

Material and methods
Study Area

The study area is located in the southwestern part of the Portuguese mainland, in the Tagus river basin, in an uneven aged cork oak woodland at the state owned farm named Companhia das Lezírias, S.A. (Fig. 1).

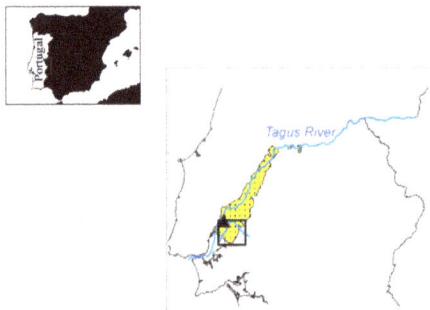

Figure 1 – Location of the state owned farm Companhia das Lezírias, S.A., at Tagus river basin, on southwestern of Mainland Portugal.

Tree Selection and Measurements

To address the stem height-cork quality relationships we have selected and measured 70-cork oak trees, previously to be harvested for their cork. These selected trees were divided in two size classes, Class 1 and Class 2, based on their stem diameter at 1.30 m height (stem diameter at breast height) over cork (dbh_{oc}). In Class 1, with the smaller trees, the trees have a $dbh_{oc} < 50$ cm; and in Class 2, with the larger trees, the trees have a $dbh_{oc} \geq 50$ cm.

In each selected tree, also previously to the cork harvesting, it was established the sunny side and shady side of the stem, for the collection of cork samples in both stem sides, at distinct heights. Afterwards, during the cork harvesting season, the diameter over (dbh_{oc}) and under (dbh_{uc}) cork was measured during the collection of the cork samples at fixed heights, at the stem base (0.3 m height) and at 1-m intervals, till the maximum harvesting height in the tree stem.

Data analysis

Linear mixed model analyses were used to test for the effects of possible influences on cork growth rates, affecting cork thickness and; on lenticels growth rates, affecting cork porosity, at different locations on the stem. Stem profiles, and correspondent cork quality, profiles were assessed at tree individual-level.

Results and Discussion

Cork thickness

Basic statistics of cork thickness variables are presented in the Table 1. The results obtained showed that there are differences on the cork thickness between trees of distinct classes. The cork growth rate was higher in larger trees (trees in Class 2, $dbh_{oc} \geq 50$ cm), with an average cork thickness of 32.1 mm (Table 1) when compared to smaller trees (trees in Class 1, $dbh_{oc} < 50$ cm), with an average cork thickness of 28.5 mm (Table 1).

These results on cork growth rates, higher in the larger trees, are in accordance with a recent study that showed that larger cork oaks grow more and significantly influence carbon cycles of cork oak woodlands, as in other forest ecosystems [12].

Table 1 – Descriptive statistics for selected cork thickness variables and for model selection. The selected cork thickness variables were: Total thickness (e_c); Annual growth rate (c_{rw}) and; Half-cycle thickness (e_{4y}).

Cork thickness variables	Acronym	Description	Tree diameter class	Mean ± Stdev	Minimum	Maximum	AIC[a]	BIC[b]
Total thickness (mm)	e_c	Commercial cork caliper (mm)	Class 1 - $dbh_{oc} < 50$ cm	28.5 ± 7.8	15.4	48.6	1013.083	1049.6809
			Class 2 - $dbhoc \geq 50$ cm	32.1 ± 7.4	16.6	49.1		
Annual growth rate (mm)	c_{rw}	Average of the 8-complete cork ring widths (mm)	Class 1 - $dbh_{oc} < 50$ cm	2.9 ± 0.9	1.3	5.2	281.289	299.588
			Class 2 - $dbhoc \geq 50$ cm	3.2 ± 0.8	1.7	5.5		
Half-cycle thickness (mm)	e_{4y}	Cumulated width for the first 4-complete cork ring widths (mm)	Class 1 - $dbh_{oc} < 50$ cm	12.9 ± 4.1	5.7	25.0	793.491	811.790
			Class 2 - $dbhoc \geq 50$ cm	14.3 ± 3.8	7.0	26.8		

[a]Akaikes's information criterion (smaller is better)

[b]Bayesian information criterion (smaller is better)

The cork thickness decreased along the stem upwards (Fig. 2). The annual growth rate of cork, c_{rw}, measured through the average of the 8-complete cork ring widths (mm), produced the lowest values of Akaikes's information criterion (AIC) and Bayesian information criterion (BIC) (Table 1). These results are in accordance with previous empirical studies that reported an overall decreasing trend of cork thickness along the stem, inherent to tree growth patterns, and associated with phellogen's intrinsic characteristics, namely the number of cork cells produced radially by the phellogen, in each cork ring [9-11] (Fig. 2).

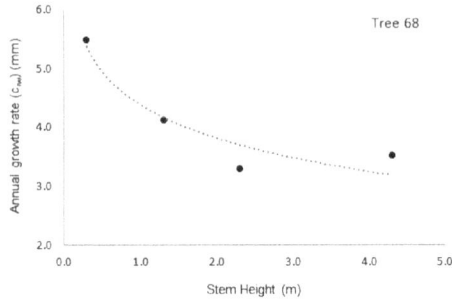

Figure 2 – Observed annual cork growth (c_{rw}) (circular black dots) with the stem height and predicted annual cork growth curve along the tree stem from the model (dashed line), for the tree no. 68.

Cork porosity
    Basic statistics of cork porosity variables are presented in the Table 2. The results obtained showed that there are differences on the cork porosity between trees of distinct classes. The cork porosity was slightly lower in larger trees (trees in Class 2, $dbh_{oc} \geq 50$ cm), with an average porosity coefficient of 9.9 % (Table 2) when compared to smaller trees (trees in Class 1, $dbh_{oc} < 50$ cm), with an average porosity coefficient of 11.3 % (Table 2).

Table 2 – Descriptive statistics for selected cork porosity variables and for model selection. The selected cork porosity variables were: Porosity Coefficient (CP); Number of pores per 100 cm^2 ($NP_{/100}$) and; Maximum pore area (A_{max}).

Cork porosity variables	Acronym	Description	Tree diameter class	Mean ± Stdev	Minimum	Maximum	AIC[a]	BIC[b]
Porosity Coefficient (%)	CP	Percentage of pore area of the total framed area	Class 1 - $dbh_{oc} < 50$ cm	11.3 ± 5.7	3.2	28.3	1820.981	1851.733
			Class 2 - dbhoc ≥ 50 cm	9.9 ± 4.6	2.9	27.2		
Number of pores per 100 cm^2	$NP_{/100}$	Total number of pores per 100 cm^2	Class 1 - $dbh_{oc} < 50$ cm	589 ± 500	216	5000	2220.426	2251.177
			Class 2 - dbhoc ≥ 50 cm	576 ± 200	268	1300		
Maximum pore area (mm^2)	A_{max}	Maximum area of pores	Class 1 - $dbh_{oc} < 50$ cm	61.7 ± 70.2	4.0	326.8	**989.643**	**1020.3950**
			Class 2 - dbhoc ≥ 50 cm	56.6 ± 98.0	3.6	756.5		

[a] Akaikes's information criterion (smaller is better)

[b] Bayesian information criterion (smaller is better)

Materials Research Forum LLC
https://doi.org/10.21741/9781644900413-1

The cork thickness decreased along the stem upwards (Fig. 3). The maximum area of pore produced the lowest values of Akaikes´s information criterion (AIC) and Bayesian information criterion (BIC) (Table 2). These results are in accordance with previous empirical studies that reported an overall decreasing trend of cork porosity along the stem, inherent to tree growth and associated with lenticular phellogen's intrinsic characteristics, namely the size of lenticels produced radially by the lenticular phellogen [9-11] (Fig. 3).

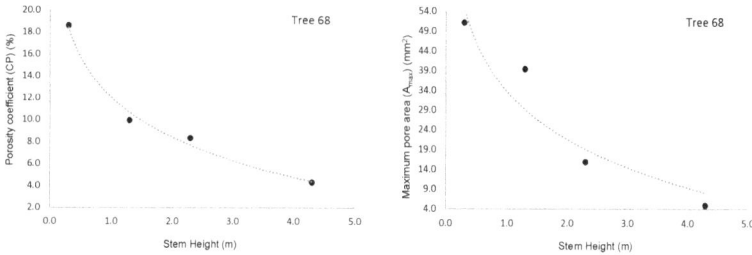

Figure 3 – Observed porosity coefficient (CP) and maximum pore area (Amax) (circular black dots) with the stem height and predicted correspondent curves along the stem from the model (dashed line), for the tree no. 68.

Stem profile
    The predicted stem profiles of cork oak trees showed distinct growth rates within the trees, along the stem upwards and; between shady and sunny sides (Fig. 4).

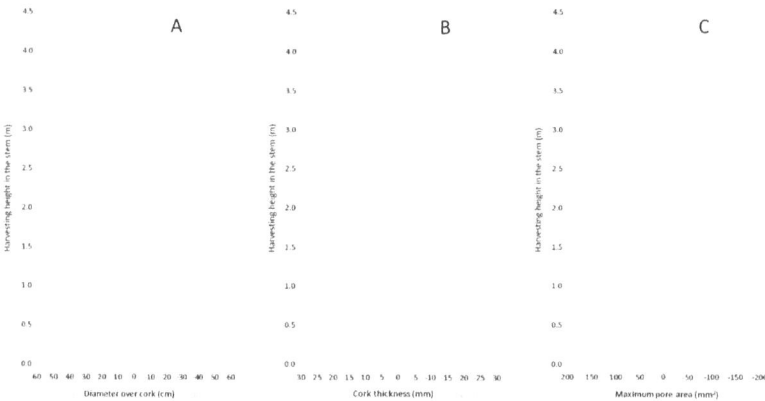

Figure 4 – Stem profiles of cork oaks showing the variations along the stem upwards and between shady and sunny sides of: (A) Diameter over cork; (B) Total cork thickness (e_c) and; (C) Maximum pore area (A_{max}).

These results on variations of radial growth patterns along the stem and also between shady and sunny sides of the stem, are in accordance with a physiological basis for enabling phellogens activities. Eventually, cork oak hormonal activity, and particularly auxin (IAA) activities, may play a relevant role on phellogen and lenticular phellogen growth patterns along the stem, as in other trees.

Acknowledgements

Project PTDC/BIA-FBT/29704/2017 *("SuberInStress - Cork formation and suberin deposition: the role of water and heat stress")*, funded by FCT/MCTES, and co-financed by FEDER in the scope of POR Lisboa 2020. Augusta Costa work was funded by FCT-MEC Post doctoral Grant (SFRH/BPD/97166/2013). Authors acknowledge the collaboration of Companhia das Lezírias, S.A. (Benavente, Portugal) for the development of this study.

References

[1] A. Costa, H. Pereira, Influence of cutting direction of cork planks on the quality and porosity characteristics of natural cork stoppers, Forest Systems 19(1) (2010) 51-60. https://doi.org/10.5424/fs/2010191-01166

[2] H. Pereira, O calibre e a porosidade como parâmetros da qualidade tecnológica da cortiça em prancha, Revista Florestal XI (2) (1998) 46-50.

[3] N. Beja, Influência da qualidade das pranchas de cortiça no rendimento e no perfil de qualidade das rolhas de cortiça natural. MSc Thesis, Tecnhical University of Lisbon , 145 pp.

[4] A. Costa, H. Pereira, Quality characterization of wine cork stoppers using computer vision, J. Int. Sci. Vigne Vin. 39(4) (2005) 209-218. https://doi.org/10.20870/oeno-one.2005.39.4.887

[5] H. Pereira, Cork: Biology, Production and Uses, first ed.. Elsevier Publications, Amsterdam, 2007. https://doi.org/10.1016/B978-044452967-1/50013-3

[6] G. Montero G, and R. Vallejo, Variacion del calibre del corcho medido a distintas alturas, Invest.Agrar. Sist. Recur. For. 1(2) (1992) 181-188.

[7] C.Taco, F. Lopes, H. Pereira, La variation dans l'arbre de l'épaisseur du liége et du dos des planches de liége pour des chênes-liéges en pleine production," Ann. Inst. Sup. Agronom. 49 (2003) 209-221.

[8] F. González, J.R. González, J.L. García de Ceca, d M. González,Variabilidad de los parametros característicos del corcho en plancha con la altura de extracción, III Congreso Forestal Español "Montes para la Sociedad del Nuevo Milenio" pp.6, 2001.

[9] J.V. Natividade, Subericultura, first ed. Ministério da Agricultura, Pescas e Alimentação, Direcção-Geral das Florestas, Lisbon, 1950.

[10] J.V. Natividade, O problema da qualidade da cortiça nos sobreirais do norte do Tejo, Boletim da Junta nacional da Cortiça 8 (1939) 5-16.

[11] D.P. Machado, Contribuição para o estudo da formação da cortiça no sobreiro, Revista Agronómica XXIII (2) (1935) 83-109.

[12] M.P. Mendes, P. Cherubini, T. Plieninger, L. Ribeiro, and A. Costa, Climate effects on stem radial growth of Quercus suber L.: does tree size matter?, Forestry 92(1) (2019) 73-84. https://doi.org/10.1093/forestry/cpy034

Cork Science and its Applications II
Materials Research Proceedings **14** (2019) 7-12

Materials Research Forum LLC
https://doi.org/10.21741/9781644900413-2

New Forest Management Techniques to Improve the Adaptation of Cork Oak Forests to Climate Change

Tusell, Josep M.[1], Mundet, Roser[2], Beltrán, Mario[3], Pique, Miriam[4], Baiges, Teresa[5], Torrell, Antoni[6]

[1] Consorci Forestal de Catalunya, Spain, josep.tusell@forestal.cat

[2] Consorci Forestal de Catalunya, Spain, roser.mundet@forestal.cat

[3] Centre de Ciència i Tecnologia Forestal de Catalunya, Spain, mario.beltran@ctfc.cat

[4] Centre de Ciència i Tecnologia Forestal de Catalunya, Spain, miriam.pique@ctfc.cat

[5] Centre de la Propietat Forestal, Spain, tbaiges@gencat.cat

[6] Forestal Catalana, SA, Spain, atorrells@gencat.cat

Keywords: Management, Adaptation, Climate Change, Forestry, *Quercus suber*

Abstract. Climate change is a serious threat to the conservation of cork oak forests, sustainable cork production and the value chain linked to this product. Mediterranean area is considered the most vulnerable bioclimatic region to climate change. The main expected impacts on the cork oak forests are [1, 2, 3]: reduction of water availability, increasing the frequency of large forest fires and more severe and more frequent episodes of pests, especially the case of cork beetle *Coraebus undatus*. In this context, new techniques and methods need to be added in the forest management, from a comprehensive approach, to improve the adaptation capacities to the climatic change of this type of forests. The Life+ SUBER project has implemented models based on the ORGEST [4]. These are irregular models with densities adjusted to the station quality with a greater proportion of large diameter trees, also achieving a high coverage of overlays that limits the development of the helioscope scrub and its vertical continuity, reducing the danger of forest fire. Depending on the station quality, they have been selective cuttings and clearing vegetation in different intensities. Of all the implemented actions a detailed technical follow-up has been carried out that has allowed to contrast its effectiveness and propose improvements to these techniques. The response of trees, in terms of growth in diameter, is always greater in managed areas than in control areas without management. At the same time, the answer is greater when selective thinning of the cork oak is done with clearing vegetation that when they are made more intense vegetation control.

Introduction

In Catalonia, data from the Forest Map of Spain (Mapa Forestal de España, DGDRPF, 2016), indicates that cork oak forests (areas where cork oak is the dominant species) occupy some 69,000 ha, of which around 29,000 ha are pure forests.

Tree density and slow growth owing to forest station conditions mean that Catalan cork is very dense, giving it a series of characteristics that are highly sought after for the manufacture of natural bottle stoppers for still wines, the cork product with the greatest added value. However, the cork is usually rather uneven and presents a higher percentage of waste than in other cork producing regions, meaning that the percentage of quality cork is usually low.

The Catalan cork present a great potential, but the reality is that cork farming and harvesting is in decline. At present, almost half of the Catalan cork oak forests are not managed. Abandonment of cork oak management is due mainly to the falling commercial value of cork, linked in turn to the decline in quality. Various factors are involved, but there is one that stands out above all the others: the already evident **effects of climate change**, particularly an increased incidence of pests and diseases – due to the presence of weaker trees – and a higher recurrence and intensity of forest fires. In Catalonia, the pest that has greatest impact on cork quality is the flathead oak borer or *Coroebus undatus*. Its larvae bore galleries in the cork, reducing the end product's quality.

These impacts will have some clearly negative effects on the productive, environmental, and social functions of cork oak forests. Furthermore, they will negatively impact the conservation status of this habitat.

Currently, sustainable forest management for cork production is one of the most important assets for cork oak forest conservation. For this reason, in this context, it is necessary to incorporate new techniques and methods in the management in order to, through an integrated approach, to improve the capacities of adaptation to climate change in this type of forest.

The Life+SUBER project

The complex problematic of cork oak forests makes necessary to address it in a multidisciplinary approach and involving different actors. Therefore, between 2014 and 2018 the Life+ SUBER project 'Integrative management for an improved adaptation of cork oak forests to climate change' was developed. In this project the following groups were represented: the private ownership (*Consorci Forestal de Catalunya*), the transformation and business industry (*Amorim*), the public administration (*Centre de la Propietat Forestal* and *Forestal Catalana*) and research sector (*Centre de Ciència i Tecnologia Forestal de Catalunya*). It has an important financial contribution of the LIFE+ programme of the European Commission. And, in addition, three public and private entities participate as co-financers: *Diputació de Barcelona, Amorim Forestal Mediterráneo SL* and *Institut Català del Suro.*

The main objective of the project is to contribute to the adaptation and an improved resilience of the European *Quercus suber* forests against the climate change, while promoting their conservation and maintaining the associated value chain.

The project is developed in four areas of Catalunya which are representatives of the different cork oak forests bioclimates: *Alt Empordà, Gavarres, Montseny- Guilleries* and *Montnegre-Corredor*. In each action area, 7 demonstratives stand have been installed (4 in a high site quality and 3 in a low site quality), where innovative management guidelines were implemented. These guidelines (which are based on the Regional guidelines and silvicultural models for sustainable forest management (ORGEST) for cork oak forests (Vericat *et al.*, 2013)) have a combined objective of improving the vitality, cork production and wildfire prevention, promoting multifunctionality through:

- **Selection thinning (irregular forest), with a variable** (although preferentially moderate) intensity, extracting up to 25% of initial basal area. The long-term objective is to generate an optimized structure for producing high quality cork, with vigorous trees with larger dimensions than the present ones and promoting regeneration in small groups rather than as single trees.
- **Selective clearings**, keeping a variable undergrowth cover ranging from 0-10% to 30-40% and respecting the species of interest for the biodiversity: strawberry tree (*Arbutus unedo), Viburnum sp.*, holm oak *(Quercus ilex) and* sporadic species. The aim is to modulate the undergrowth cover and height according to each stand, to promote horizontal precipitation capturing, reduce the structural vulnerability to crown fire, keeping auxiliary fauna (predators of *Coraebus)* and achieve an adequate cost-efficiency.
- **Logging residues treatment** to reduce fire risk; residues are treated *in situ* to facilitate their decomposition or are extracted and chipped besides forest roads.

Cork Science and its Applications II   Materials Research Forum LLC
Materials Research Proceedings **14** (2019) 7-12   https://doi.org/10.21741/9781644900413-2

Figure 1. Aspect of two stands before (left half of the picture) and after (right half) the silvicultural interventions

Also, during the Life+ SUBER project, several actions were done in order to improve the control of *Coreabus undatus* plague. These actions are related to the massive trapping and always in line with other actions of biorational control. In total, we installed 720 traps distributed within the cork oak forests suffering a higher incidence of this pest during 2015, 2016 y 2017.

Results in terms of vitality and productivity
The density of small trees is reduced, while largest trees are kept in the forest. After the interventions, the small trees grow more and changes in diameter classes are observed, such as increase in the proportion of trees of the medium-sized group. Thus, the stands tend towards the pre-established management guidelines.

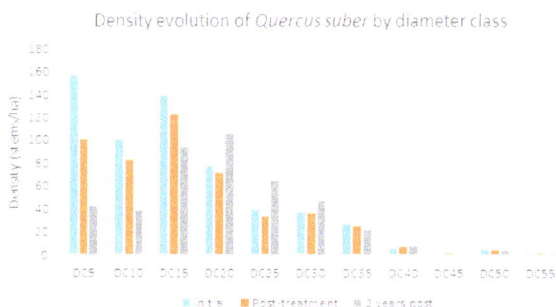

Figure 2. Quercus suber diameter distribution before, after and two years after the silvicultural treatments. Summary of results of 16 stands (Action B1).

An intensive monitoring of the intervened stands was done to improve the knowledge transfer to stakeholders. It is expected to keep on conducting further monitoring in the future to characterize the long-term effects of the demonstrated silviculture. In any case, two years after treatments we could detect some differences in the vitality response of the trees:

- The selective felling combined with a low intensity selective shrub clearing led to a relative increase in diameter growth ranging from 7% (high quality sites) to 21% (low quality sites), while control areas showed no significant growth.
- The same selective felling combined with a more intense shrub clearing sled to a relative increase in diameter growth ranging from 2% (high quality sites) to 8% (low quality sites), while control areas showed no significant grow.

Therefore, the stands where the selective felling was combined with a soft selective shrub clearing showed a better tree growth than the stands subject to more intense clearings, particularly in low quality sites. The interaction of shrubs and undergrowth with the water available for trees seems to be significant in relation with the growth and vitality of cork oak forests.

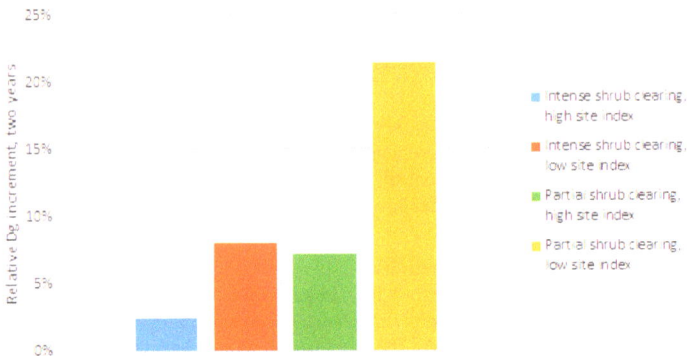

Figure 3. Summary of relative growth increase of average diameter in stands subject to diferent treatments among the demonstrative stands of Life+ SUBER project.

Results in terms of large wildfire prevention
An improvement in large wildfire prevention is observed due to a decrease in the vulnerability of forest structures against high intensity crown fires, especially in the stands where this objective was set as prioritary. The forest structures before the silvicultural treatments showed in general a high vulnerability to crown fires, those that generate high intensity wildfires. There is an immediate improvement after the treatments, which in general remains perceptible over the following two years.

Figure 4. Demonstrative stand of Life+ SUBER project where the treatment done consisted on the creation of open forest areas (Dehesa-type), being wildfire prevention the main objective.

Therefore, the adaptative management strategy should combine the cork production with fire prevention through a landscape-scale Planning. The recommendations are focused in the assignment of fire prevention as preferent objective in strategic areas with regard to fire behavior, while in other areas the increment of cork production (always considering sustainability and multifunctionality) is set as preferential objective.

Results in terms of biorational control of *Coraebus undatus*
The knowledge about the insect biology has progressed notably during the years of execution of the Life+ SUBER project, and the techniques demonstrated in it proved to contribute significantly. In any case, there are still many aspects of the biology of this species that remain unknown, so it is necessary to continue conducting experiences related with the biorational control of this plague. Some conclusions of the Life+ SUBER project experience are:

- The best trapping results of adult insects were obtained with the purple triangular prism trap, using 8 traps/ha, at 170 cm from ground, using a diffuser of luring substances and clearing the shrubs in about 25 m^2 and keeping the trap free of foliage.
- The number of captures is very variable from one year to another. A great variety of implied factors exists, including the biology and population dynamics of the insect and the meteorology of previous years. **A possible relation between the low precipitation of the year before the laying and the increase of the presence of the plague have been suggested.**
- If the trapping is able to reduce in an effective way the insect population and the effect of the undergrowth clearing on the plague incidence cannot be determined with precision. However, the knowledge of the insect biology has been improved, in terms of behavior, distance and flying season.

All the more detailed information of the results of the project can be found in the guide 'Climate change measures and recommendations for cork oak forests' [5]

References

[1] Pereira, J. S.; Vaz Correia, A.; Joffre, R. 2009. Facing climate change. En: Aronson, J.; Pereira, J. S.; Pausas, J. D. (eds.). Cork oak woodlands on the edge: ecology, adaptive management, and restoration. Island Press. Washington, DC, pp. 219- 226.

[2] Díaz, M.; Pulido, F. J.; Pausas, J. D. 2009. "9330 Alcornocales de *Quercus suber*". En: VV.AA. Bases ecológicas preliminares para la conservación de los tipos de hábitat de interés comunitario en España. Ministerio de Medio Ambiente, y Medio Rural y Marino. Madrid.

[3] Vericat, P., Piqué, M., 2012. El cambio global: impactos probables sobre las formaciones de *Quercus* y gestión para la adaptación. En: Vericat, P.; Piqué, M.; Serrada, R. (eds.). Gestión adaptativa al cambio global en masas de *Quercus* mediterráneos. Centre Tecnològic Forestal de Catalunya. Solsona (Lleida), pp. 29-46.

[4] Vericat, P.; Beltrán, M.; Piqué, M.; Cervera, T. 2013. Models de gestió per als boscos de surera: producció de suro i prevenció d'incendis forestals. Sèrie: Orientacions de gestió forestal sostenible per a Catalunya (ORGEST). Centre de la Propietat Forestal. Departament d'Agricultura, Ramaderia, Pesca, Alimentació i Medi Natural. Generalitat de Catalunya.

[5] Mundet, R.; Baiges, T.; Beltrán, M.; Torrell, A. 2018. Climate change measures and recommendations for cork oak forests. Life+Suber Project

Cork: New uses in Architecture

C. Verissimo

Dalhousie University, NS, Canada

Abstract. This research will create greater bring awareness of the potentialities of cork as sustainable material and seek possible new uses of cork materials in architecture. Cork is a natural, recyclable, renewable and non-toxic resource, with strong impact in Mediterranean culture and ecosystems for centuries. This research provides an environment for researchers and architects as well as those in the industry and other interested parties to meet and develop ideas and experiments using cork. The results proved themselves to be beyond expectations.

Cork Oak forests and Culture

Cork is a natural, recyclable, renewable and non-toxic resource, harvested from the living bark of the Cork Oak, Quercus Suber L. (Pereira, 2007)

Though the Cork Oak can flourish in many climates, the conditions that favour commercial use lie in a fairly narrow swath that cuts through Western Europe and Northern Africa along the Mediterranean coast. Cork Oak forests are an agro-silvo-pastoral ecosystem, built by humans and preserved by control of shrubs and using natural resources (Potes, J. & Babo, H. 2003).

Figure 1. agro-silvo-pastoral ecosystem.

They are maintained through the persistence, hard work and passion of generations that for centuries have planted Cork Oak trees for the eventual benefit of their grandchildren, who profit from its harvesting. Recent developments in cork research demonstrate that its application goes far beyond the classical cork-wine cultural relationship that is globally recognized. (Gil., 2007)

Cork Science and its Applications II   Materials Research Forum LLC
Materials Research Proceedings **14** (2019) 13-19   https://doi.org/10.21741/9781644900413-3

Cork and ARCHITECTURE

The use of cork is ancient, especially along the Mediterranean with its use from daily life objects to architecture. Existing buildings such as the Convent of the Capuchos of the XVI century, Sintra, Portugal, still shows how cork was used as a finishing and outdoor material for comfort (Muchagato J.; Oliveira Martins A. 2013). Cork was an early product in a global market due to Portuguese and Spanish empires of the XVI century (Santos, C.O., 2000).

Figure 2 – Capuchos Convent, Portugal by JBond (left). Detail of a Front door of a House in Monsaraz, Portugal (right).

Cork is very versatile and adopts different technological transformation processes, giving rise to several products, which can be used in different applications. New demands for cork give rise to new technologies that can replace traditional methods of boiling the cork, through a microwave process that maintains cork qualities but expands the material's capacity to three times more than its original size.

Cork products for the construction industry are most suitable for sustainable and efficient energy construction, given its mentioned ecological characteristics. In addition, these products contribute to general comfort and indoor air quality (Gil, 2007).

This new technologic development gives an opportunity in architecture for the use of cork as a primary material as in the past. In the last decade some examples of the use of cork have emerged, such as: Portugal Pavilion for Expo 2000 Hannover exhibition, Serpentine Pavilion, London, 2012 or Portugal Pavilion Shanghai Expo 2010.

Figure 3. Alvaro Siza, Eduardo Souto de Moura, Portugal Pavilion for Hannover, Expo 2000 (left). Herzog & DeMeuron and Ai Weiwei's cork Serpentine Gallery Pavilion 2012 (right).

Research and Methodology

This research started July 2014, with the organization of an international workshop in Lisbon, focused on the use of cork as a material in architecture.

Materials Research Forum LLC
https://doi.org/10.21741/9781644900413-3

It provided an environment for researchers and architects as well as those in the industry and other interested parties to meet and develop ideas and experiments using cork.

The research proposes different challenges of the use of cork as a building envelope material. It is organized by themes, according to the type of materials/prototypes that were tested or developed.

TOPIC 1: Natural Cork - Tradition

The topic was the use of Natural Cork – Tradition. Natural cork had multiple applications in architecture, from cladding to isolation. The objective of this topic is to understand how some of those examples can be adapted for use in construction. Therefore, the research concentrated: 1 – To gain an appreciation and understanding of Cork as a material. 2- To understand traditional methods in the use of Cork and learn from that experience. 3 – To Conduct explorations with natural cork or products using only natural cork, to understand its potentialities and possible applications as a building envelope.

Extended research was conducted into recycling of residuals of natural cork produced by cork stopper manufacture. Since cork is a material with enormous potentialities of application, to reinforce the methodology, we divided the research into three main subjects for the workshops: cork as a membrane, cork as a finishing and cork as self-supporting material.

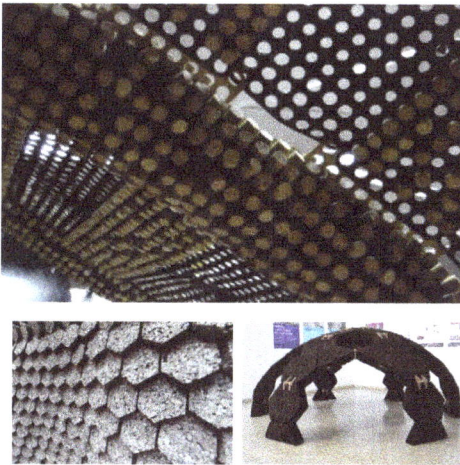

Figure 4. Cork used as a membrane (top). Cork used as a finishing material (left). Cork used as a self-supporting material (right).

These are three examples, of the over 20 experiences developed: in figure 4, cork was used as a membrane. The incorporation of recycled cork stoppers gives a surprising structure to the fabric and added thermal insulation. The black cork agglomerated that was shaped into a hexagon tile that was applied as a finishing material to a curved surface, with thermal and acoustic benefits (left). On the right of figure 4 a dome was assembled using natural cork for joints and cork natural black agglomerated. This dome can be folded and transported due to its lightweight.

Cork Science and its Applications II   Materials Research Forum LLC
Materials Research Proceedings 14 (2019) 13-19   https://doi.org/10.21741/9781644900413-3

TOPIC 2: Cork Composite – Industry

The topic was the use of Cork Composite – Industry. On the last decades there has been an important effort, from the industry, to develop several new derivative products of cork incorporating new technologies applied to traditional products.

The objective of this topic is to understand how some of these materials or products will be adapted for use in construction. Therefore, this topic research concentrates on: 1 - to know the industrial cork processing chain and the composite cork products. 2 - Selection and characterization of cork products and their use changes, i.e. with aging and weathering, with a selection of performance-related indicators. 3 - Conduct experiences in the Workshop involving the industry, overcoming potentialities of their materials.

Extended research was conducted into recycling of residuals of composite cork produced by cork industry. It was used the same methodology as in previous year: cork as a membrane, cork as a finishing and cork as self-supporting material.

Figure 5. Cork used as a membrane (top). Cork used as a finishing material (left). Cork used as a self-supporting material (right).

These are three examples, of the over 20 experiences developed. In figure 5 (top), cork was used as a membrane. The double layer triangle pattern cork fabric is glue to natural fibres that work as a connecter element for the pattern. The interior of the triangles is filled with cork granules adding thermal insulation. As alternative the triangles can be insufflating with air. In figure 5 (left), Rubber Cork Composite and Cork Resins composite materials, creating a tile through CAD-CAM process. A pattern was carved in the titles and flied with natural resin for transparency. The tiles were attached, as a finishing material, to metal structure, with thermal and acoustic benefits. In figure 5 (right) a small pavilion was assembled using laminated wood and cork for surfaces and black rubber cork for joints. The pavilion was assembled without any metal joints.

TOPIC 3: New Materials - Future Technologies
The topic was New Materials - Future Technologies. In the last decades there has been an important awareness that architects must start looking at natural and sustainable materials for building construction.

Therefore, this topic is focus on: 1 - Understand current architectural practices and foreseeing their future needs; 2 - Selection of new materials and technologies that can in the future bring new possibilities of use in architecture and product design; 3 - Understand how research and industry can work together to develop strategic possibilities for future applications in architecture.

These three examples show the final pieces developed from over 20 examples. In Figure 6 (top), a wooden structure is holding layers of cork composite material, which were parameterized using computer software to create grading openings from opaque to more transparent. A three-dimensional butterfly-joint was designed and cast using a CNC machine. The different butterfly-joints with angles of 30°, 45°, 60° and 90° allowed for multiple possibilities of assemblage. The strength of this high-density composite cork agglomerated allowed for different types of assemblage using these joints. It creates a system that can generate different shapes using only cork materials to structure and support. (left). A small tent was developed using a cork textile and an origami technique. The folding of the origami creates stiffness to the textile and allows for an easy folding of the tent (right).

Figure 6. Cork is used as a cladding material (top). Cork used as a self-supporting material (left). Cork used as a membrane (right).

Research dissemination

For this research we selected 30 participants per topic, from Portugal, other European countries, Brazil, Canada and Japan. Five Portuguese Universities and one Canadian University were involved. We also had the participation and contribution of well recognized researchers in cork materials, engineering and biology.

The involvement with the industry is crucial for this research. This research showed that it is possible to work at the same time with different cork companies with a large spectrum of materials. On the 27th of May of 2016, the Observatorio da Cortiça (Cork Observatory), organized a conference "Cork - News Uses in Architecture", in partnership with the University of Lisbon and Dalhousie University. That conference put together industry, universities, architects and designers, discussing the future opportunities the use of cork materials in architecture and design.

A Facebook page was created in 2014. [https://www.facebook.com/corkworkshop, accessed on 02/Jul2019 19:51].

Since 2014, an itinerary exhibition has been showing the work of this research across Portugal: University of Lisbon Architecture School, Lisbon 2014-16, Observatorio da Cortiça in Coruche 2015-2019, Feira do Montado de Portel, 2016, Ponte de Sor / Montargil 2018, J.S. Cork in Lisbon 2019.

Figure 7. View of the exhibition, at the Observatorio da Cortiça in Coruche, 2018

Conclusions

This research creates awareness on the potentialities of cork as a sustainable material and seek possible new uses of cork materials in architecture.

As a result, it contributes to a greater sustainability in the construction industry and opportunities of market in the cork sector.

From these three-year initiatives' results, there is a clear interest by the different Universities, institutions and Industries to continue to support this research and the work developed in the workshops. It has been demonstrated that it is a wonderful platform to reach out to young generations of future architects and designers about the opportunities to use cork in architecture. This research explored the cork industry wastage, cork recycling and new cork-based materials, which are still in various stages of development with enormous potential for the construction, related industries.

We tested and adapted materials developed for industrial uses into possible architectural uses, hoping that in the future there will be greater application in architecture and eventually will contribute to a greater sustainability in the construction and design industries as well as in the cork sector.

References

[1] Pereira H., 2007. "Cork: Biology, Production and Uses". Amesterdam: Elsevier: 26-29, 33 -53. https://doi.org/10.1016/B978-044452967-1/50004-2

[2] Potes, J. & Babo, H. 2003. "Montado" an old system in the new millennium" African Journal of Range & Forage Science, 20 (2): 131-146. https://doi.org/10.2989/10220110309485808

[3] Gil, L., 2007 "Cork as a Building Material. Technical Manual". Santa Maria de Lamas: APCOR.

[4] Gil L. 2010. "A cortiça, o ambiente e a sustentabilidade" (in Portuguese). Biol. Soc. 10, 13–15

[5] Gil L. 2011. "Environmental, sustainability and ecological aspects of cork products for building". Sci. Technol. Mater. 23, 87–90

[6] Potes, J. & Babo, H. 2003. "Montado" an old system in the new millennium" African Journal of Range & Forage Science, 20 (2): 131-146. https://doi.org/10.2989/10220110309485808

[7] Santos, C.O., 2000 (in Portuguese) "O Livro da Cortiça". Ed. Carlos Oliveira Santos/Diglivros. (65-72)

[8] Santos, C. O. & Amorim, A., 2008, "Clusters United by Nature: The World of Wine and Cork", Amorim Group, 29-32

[9] Muchagato J.; Oliveira Martins A., 2013" Convento of the Capuchos", Parques de Sintra and Scala Arts & Heritage Publishers,

[10] Pestana, M. e Tinoco, I., 2008 (in Portuguese) "A Indústria e o Comércio da Cortiça em Portugal durante o Século XX", 1-5

[11] Gil L., Moiteiro C. (2003). "Cork" in Ullmann's Encyclopedia of Chemical Technology", 6th Edn. Germany: Wiley-VCH Verlag. https://doi.org/10.1002/14356007.f07_f01

[12] Chebao, F. 2011, (in Portuguese) "Cortiça e Arquitetura". Euronatura, 51-81

Materials Research Forum LLC
https://doi.org/10.21741/9781644900413-4

Behavior of Natural Cork Stoppers when Modifying Standard Corking Parameters: Three Practical Cases

Prades López, C.[1,a,*] and Sánchez-González, M.[2,b]

[1]Departamento de Ingeniería Forestal – Universidad de Córdoba Campus de Rabanales. Edificio Leonardo Da Vinci, Ctra. Madrid, Km 396 - Córdoba, Spain

[2]Centro de Investigación Forestal – CIFOR. Instituto Nacional de Investigación y Tecnología Agraria y Alimentaria – INIA. Ctra. De la Coruña, km 7,5 - 28040 Madrid, Spain

[a]ir1prloc@uco.es, [b]msanchez@inia.es

Abstract. In the elaboration of wine many factors directly affect the quality and properties of the final product, among them the choice of bottle and stopper. Standard cork stoppers measure 24 mm in diameter. This dimension determines the thickness or minimum caliper of the cork on the tree, which should be approximately 29 mm for the manufacture of one-piece natural cork stoppers. In Spain, an average of 54,614 tons of cork were produced per year in the period 2006–2013. However, both the thickness and quality of the cork has decreased, thus affecting the percentage of cork that can be used to manufacture natural stoppers, as well as the quality of the stoppers produced. This declining trend could be stabilized or reversed when new cork plantations enter into production following the reforestation of agricultural land. This work aims to address a current need that has arisen in the sector: to increase the percentage of cork stoppers of sufficient caliper and quality for the manufacture of one-piece natural stoppers. In order to increase the quantity of cork suitable for manufacturing natural stoppers, it is necessary to modify the corking diameters by reducing the diameter of the stopper and the compression rate, while ensuring the impermeability of the cork to liquids and gases.

Keywords: Cork, Bottleneck Diameter, Stopper Diameter, Compression Rate, Compression Force, Relaxation Force, Diametrical Recovery

Introduction

In the elaboration of wine, there are several factors that directly affect the quality and properties of the final product, among them the choice of bottle and the choice of stopper. One-piece natural stoppers are the highest value-added product manufactured from high-quality cork. The physical and mechanical properties of this natural product make it ideal for sealing fine wines. Standard stoppers are cylindrical in shape, weigh 4 grams and measure 44 mm in length and 24 mm in diameter. It is this last dimension that determines if the cork on the tree is of the sufficient thickness or caliper for the manufacture of one-piece stoppers as the cork bark must have a minimum thickness of approximately 29 mm.

Bottling is the last phase of the winemaking process in which the winemaker intervenes. This is a very important phase as the evolution of the bottled wine will depend on the characteristics of the stopper and the bottling practice. The main physical and mechanical parameters of the cork that influence the sealing capacity of stoppers are the density (a parameter of quality), the compression force required to reduce the diameter of the cork to the diameter of the corking jaw, the relaxation force the cork exerts on the bottleneck on insertion, the recovery of the cork diameter and the extraction force necessary for the final consumer to remove the stopper (González-Hernandez et al., 2014).

To seal wine bottles, the stopper is compressed in the corking jaw and inserted into the bottleneck. In normal practice, there is a close relationship between the diameter of the bottleneck,

Materials Research Forum LLC
https://doi.org/10.21741/9781644900413-4

the diameter of the stopper and the diameter of the corking jaw. Standard bottlenecks are not completely cylindrical but of a conical, tapered shape. According to European Standard EN 12726 (2000), standard bottlenecks range in diameter and height, from 18.5 ± 0.5 mm (diameter) at 3 mm (height), 19 mm (average diameter) at 22.5 mm, to 20 ± 1 mm (diameter) at 45 mm (height). The optimal stopper diameter for still wines corresponding to a standard 19 mm bottleneck is 24 x 44 mm (figure 1).

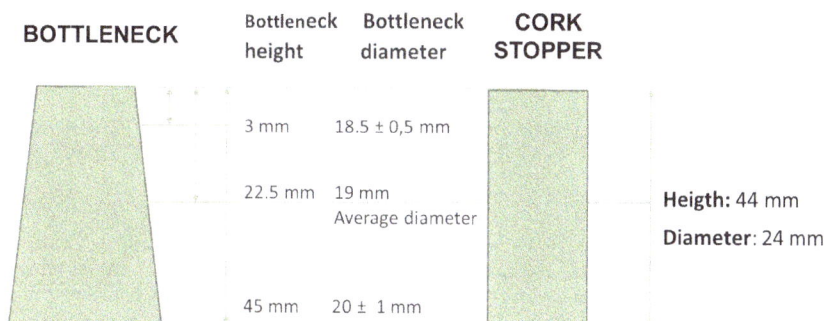

Figure 1. Bottleneck and cork stopper dimensions for still wines

The stopper and jaw diameter are related through the compression rate. In usual bottling practice, corking machines reduce the stopper diameter by 33% (33% compression rate) from 24 to 16 mm in order to insert the stopper into 19 mm bottlenecks.

A total of 201,000 tons of cork are currently produced per year worldwide, of which 62,000 tons are produced in Spain (Cork Quality Council, 2015). The average cork production in Spain has decreased from 74,500 to 66,925 and 54,614 tons per year in the periods 1951–1960, 1961–1970 and 2006–2013, respectively (MAPAMA, 1940–1971; MAPAMA 2005–2013). The decrease in cork production refers to both the cork thickness and quality, which has affected the percentage of cork that can be used for manufacturing natural stoppers and the quality of the stoppers obtained.

However, this declining trend could stabilize or be reversed. The entry into production of new cork plantations under the framework of the reforestation plan of agricultural lands promoted by the European Union (Council Regulation (EEC) 2080/92 of the Council of 30 June 1992) is expected to increase cork production. In Spain, 83,425 ha. were reforested with cork oak in pure and mixed masses in the period 1993–2000 (Ovando et al., 2007).

Climate change is another important factor in the long-term production of cork. The influence of climatic variables on cork growth has been widely documented, with very significant correlations found between production and droughts and the precipitation regime (Ghalem et al., 2018). Recognizing these factors can improve decision- making processes and reduce the negative effects of climate change (Rodney, 2015).

Given these circumstances, cork of a smaller thickness or caliper could be considered for the manufacture of one-piece natural cork stoppers. This could increase the amount of cork suitable for natural stoppers, provided that the sealing capacity of the cork is maintained or improved.

From a mechanical viewpoint, it is possible to improve sealing quality in three ways. Firstly, by increasing the stopper diameter; a common practice in wineries for the bottling of long-aging wines.

Secondly, by selecting higher quality stoppers, which would increase the cork's capacity for diametrical recovery. And, thirdly, by decreasing the compression rate, which would also increase the capacity for diametrical recovery.

Based on the hypothesis that cork production follows a declining trend, the first two options must be ruled out. Therefore, in order to increase the quantity of cork suitable for manufacturing natural stoppers, it would be necessary to modify the diameters involved in the corking process, reducing the stopper diameter and the compression rate to preserve the quality and properties of the wine.

This study attempts to address a current need in the cork sector: to increase the percentage of cork of sufficient caliper and quality for manufacturing one-piece natural stoppers. The aim is to evaluate the mechanical behaviour of stoppers during the corking process in three practical cases by reducing the stopper diameter, reducing the bottleneck diameter and reducing the diameter of the corking jaw. This work can be considered a previous step towards establishing the technical criteria to optimize the correlation between corking diameters and the forces that intervene in the corking process.

Methodology

The main objective is to evaluate corking performance when reducing the stopper diameter. In order to ensure that stoppers of a smaller diameter maintain the proper sealing conditions, the diameter of the bottleneck and the compression rate must also be modified. For all three cases, tests will be carried out to determine the effect of the bottleneck diameter, the stopper diameter and the jaw diameter on the corking process, as described below.

As an initial hypothesis, we assume three bottleneck diameters: standard bottlenecks of 19 mm in diameter and smaller bottlenecks of 18 and 17 mm in diameter (Table 1).

To determine the stopper diameter (SD) from the bottleneck diameter (BD), the simplest approach is to assume a linear relationship (SD = BD +5) or a proportional relationship (24.BD = 19.SD), which are obtained based on a bottleneck of 19 cm in diameter corresponding to the optimal diameter of a 24 cm stopper. However, these relationships cannot be extrapolated. Therefore, in the absence of an established criterion, stopper diameters of 24 mm, 22.5 mm and 20.5 mm for bottleneck diameters of 19 mm, 18 mm and 17 mm, respectively, will be used (Table 1).

Stopper deformation (ε) is directly related to the difference between the stopper diameter (SD) and the bottleneck diameter (BD) [$\varepsilon = 1 - (SD/BD)$] (Pereira, 2007). To ensure correct bottling, a compression rate higher than 33% should not be applied due to the negative effect it has on the elasticity, dimensional recovery and relaxation force of the cork. These negative effects could be offset by increasing the stopper diameter. Similarly, a decrease in the stopper diameter could be compensated by decreasing the compression rate. In order to not over- compress stoppers of a smaller diameter, in this work the compression rate is not estimated as a percentage of the stopper diameter, but from a linear relationship between the jaw diameter (JD) and the bottleneck diameter (BD):

$$JD \text{ (mm)} = BD - 1.5 \text{ mm}$$

Therefore, a compression rate of $\approx 27\%$ should used for the 24 mm and 22.5 mm stoppers and a compression rate of $\approx 25\%$ should be used for the 20.5 mm stoppers (Table 1).

Cork Science and its Applications II   Materials Research Forum LLC
Materials Research Proceedings **14** (2019) 20-27   https://doi.org/10.21741/9781644900413-4

Table 1. Corking diameters and compression rates used in the three pracical cases

Parameters	Practical cases		
Number of stoppers	30	30	30
Bottleneck diameter BD (mm)	19	18	17
Stopper diameter SD (mm)	24	22.5	20.5
Jaw diameter JD (mm)	17.5	16.5	15.5
Compression rate (%)	27	27	25

The experimental material consists of three batches of 30 one-piece natural stoppers of the same quality (Class 1) measuring 44 mm in length and 24, 22.5 and 20.5 mm in diameter for 19, 18 and 17 mm bottlenecks, respectively. Tests were carried out on each batch of 30 stoppers to simulate the corking process and determine the diameters and forces that intervene in the process.

- ✓ Diameter of cylindrical stopper (SD) (mm)
- ✓ Recovered diameter (RD) (mm): diameter measured 24 h after the compression force has ceased (maximum value of the recovered diameter if the compression force has ceased and the stopper is not inserted into the bottleneck)
- ✓ Maximum compression force (CF) (daN): The compression force of the stopper is defined as the radial and perpendicular force that the corking jaws exert on the lateral surface of the stopper, reducing its diameter by a percentage equivalent to the compression rate.
- ✓ Maximum relaxation force (RF) (daN): The relaxation force is defined as the radial and perpendicular force exerted by the cork stopper on the inner walls of the bottleneck. The maximum relaxation force is exerted by the stopper on the bottleneck on insertion.
- ✓ Relaxation ratio (RR) (%)
- ✓ Elastic recovery or diametrical recovery of the stoppers (DR) (%)

The stoppers were acclimated at the INIA-CIFOR Cork Laboratory for a period of 30 days at a temperature of 20°C and a relative humidity of 65% where they acquired a moisture content of approximately 6%. Once acclimated, the diameter of the stoppers was measured in mm and the tests were carried out.

Compression Test: The CF (daN) was measured using a corking machine equipped with a load cell (UTILCELL. Model: 650 SNo 460775(02) Emax: 2Tn). The closure diameter of the jaw was selected and the compression and insertion force of the stopper in the bottleneck was measured. The maximum CF was measured and recorded for each stopper (González-Hernández et al., 2014).

CF (daN) = maximum compression force

Relaxation Test: To measure the RF (daN), the stopper was placed in a bottleneck tube and inserted into a device developed at the INIA-CIFOR Cork Laboratory (Gonzalez Hernandez et al., 2012). The RF exerted by the stopper against the inner walls of the tube is recorded and transmitted via a spindle to a load cell (SENSOCAR, Mod. S-1, Emax. 150 kg. Precision 50 g). The load cell then transforms the RF into an electrical signal which is shown as a value. The fitted stopper remained in the device for 30 minutes and the maximum RF of each stopper was recorded.

RF (daN) = maximum relaxation force

Relaxation Ratio (RR, %): The action of external forces on a deformable solid produces energy which is stored in the form of potential energy, thus increasing the internal energy. In an elastic solid, this process is reversible and there is no loss of energy when the effort ceases (Ortiz Berrocal, 2011). The viscoelastic behavior of the fitted stopper is measured by the relaxation ratio (Gonzalez-Hernández et al., 2014):

$$RR(\%) = 100 \frac{RF}{CF}$$

where CF (daN) is the maximum compression force and RF (daN) is the maximum relaxation force.

Elastic Recovery or Diametrical Recovery (DR, %): If the stopper is released from the corking jaw and not inserted into the bottleneck, the stopper recovers its diameter until reaching the maximum value. Twenty-four hours after compression, the DR of the stopper is measured at the same point where the initial diameter of the stopper was measured (SD). DR is calculated as:

$$DR(\%) = \frac{SD}{RD} 100$$

where SD (mm) is the stopper diameter and RD (mm) is the maximum recovered diameter after compression.

It is assumed that effective closure is achieved when the relationship between the maximum compression force exerted during bottling and the force exerted by the fitted stopper on the bottleneck remain constant, that is, when the stoppers have the same relaxation coefficient.

Results and discussion
The mean CF values were 196.34 daN, 188.59 daN and 172.37 daN for the 24 mm, 22.45 mm and 20.5 mm stoppers, respectively (Table 2), thus indicating that the CF decreases as the compression rate decreases. The values obtained for the 24 mm stoppers at a compression rate of 27% were lower than those obtained by González- Hernández et al. (2014) and Prades et al. (2014), which were around 230 daN, for stoppers of the same diameter and a compression rate of 33%.

For standard bottling values, the compression stress (σc) applied to the surface contact area in the radial-tangential direction (σc = E, ε) (Pereira, 2007) takes values of $\approx$ 1 MPa when the stopper is compressed in the corking jaw. The compression stress obtained for the three batches of stoppers are somewhat lower and range from 0.8 to 0.83 MPa, and remain in the same region of the curve corresponding to the densification of the cell.

Materials Research Forum LLC
https://doi.org/10.21741/9781644900413-4

Table 2. Characterization of stoppers: mean (standard deviation shown in parenthesis) and range for the three batches

Variable		STOPPER DIAMETER					
		24 x 44		22.5 x 44		20.5 x 44	
	n	Mean	Range	Mean	Range	Mean	Range
SD (mm)	30	24.36 (0.07)	24.19- 24.48	22.7 (0.06)	22.57- 22.86	20.70 (0.05)	20.62- 20.81
CF (daN)	30	196.34 (26.3)	116.6- 246.0	188.59 (20.85)	145.6- 232.2	172.37 (25.03)	124.20- 211
RF (daN)	30	25.72 (3.28)	17.93- 32.68	22.35 (2.5)	14.75- 26.36	18.98 (2.21)	14.16- 24.60
DR (%)	30	0.97 (0.01)	0.96-0.98	0.97 (0.01)	0.96-0.99	0.97 (0.01)	0.96-0.99
RR	30	0.13 (0.01)	0.10-0.15	0.12 (0.01)	0.09-0.14	0.11 (0.01)	0.09-0.13

The relaxation force was measured with the device patented by the INIA-CIFOR Cork Laboratory (González-Hernández et al., 2012) and showed mean values of 25.72 daN, 22.35 daN and 18.98 daN (Table 2), which are somewhat lower than those obtained for standard sealing conditions. The relaxation force is not usually measured directly. Other authors (Fortes et al., 2004) have reported values ranging from 40 to 70 daN. These differences are because the test used to determine the relaxation force was different from the one used here.

The relaxation ratio values range from 1 when the compression force is equal to the relaxation force (bottleneck diameter = jaw diameter) to 0 when the relaxation force is equal to zero (bottleneck diameter = maximum recovered diameter). The relaxation ratio was 0.13, 0.12 and 0.11 for the 24 mm, 22.5 mm and 20.5 mm cork stoppers, respectively (Table 2). For standard sealing conditions, the relaxation ratio takes values of 0.11. Giunchi et al. (2008) define the *resilience index* as the ratio between the relaxation area and the compression area of the stress-strain curve and obtained values of 0.24–0.29 for cork stoppers. However, although the concepts are similar, the methodologies are different, so the data are not comparable.

The capacity of the stopper to recover its diameter, which is related to the relaxation force, decreases with time, thus affecting the sealing capacity and impermeability of the bottle closure. The average values are 96% under standard sealing conditions. In the three cases tested, the stoppers recovered 97% of their diameter. This is one point higher than standard recovery due to the decrease in the compression rate (Table 2). Elastic recovery is very important in stopper mechanics, and can be considered a measure of the sealing capacity and impermeability of the stopper. Theoretically, a DR of 97% ensures correct sealing conditions for the new diameters (bottleneck and stopper). However, it is not possible to conclude whether the relationship between the diameters corresponds to the optimal situation.

Conclusions
The compression and relaxation forces are directly related to the compression rate, and both are found to decrease as the compression rate and diameter decrease. For a compression rate of 27% and 24 mm and 22.5 mm stoppers, the compression force decreases from 196.34 to 188.59 daN, respectively, and the relaxation force from 25.72 to 22.35 daN, respectively. For a 25% compression rate and 20.5 mm stoppers, the compression force is 172.37 daN and the relaxation

force is 18.98 daN. A similar relaxation ratio ranging from 0.13, 0.12 to 0.11 was obtained in all three cases.

The values obtained for the compression force and relaxation force are somewhat lower than those obtained for standard sealing conditions. The compression stresses are also somewhat lower than the mean stress under standard conditions, but they remain in the same region of the curve corresponding to the densification of the cell.

The results of the tests with bottleneck diameters of 19, 18 and 17 mm and compression rates equivalent to BD - 1.5 mm were satisfactory, as a diametrical recovery of 97% was obtained. This is higher than the diameter recovered in standard conditions, thus ensuring the impermeability of the closure. However, it is not possible to conclude whether the ratio between the tested diameters is optimal.

The quality of the cork stoppers has not been considered in this work. However, due to the heterogeneity of the material (Anjos et al., 2008), it would be of interest to study the influence of cork quality and density in the relationship between the diameters and forces involved in the corking process.

If it is possible to reduce the diameter of stoppers, the impact of the new raw material requirements regarding cork production and the corking process should be assessed. In particular, parameters such as corkage shifts and the debarking height depend to a large extent on the caliper of the cork.

In modifying the diameter of cork stoppers, the diameter of the bottlenecks and the compression rate must also be modified. The optimal relationship between stopper diameter and bottleneck diameter should be established based on the mechanical behaviour of natural cork stoppers during the corking process, taking into account the compression (deformation) and relaxation (recovery) forces exerted on the stoppers. This study can be considered a first step in that direction.

Acknowledgements

This study has been carried out within the framework of the CC13-045 collaboration agreement between INIA-CIFOR and the University of Cordoba, Spain. The authors would like to thank the laboratory technicians Maria Luisa Cáceres Esteban and Lorenzo Ortiz Buiza for their contribution to the work.

References

[1] Anjos O, Pereira H, Rosa ME. 2008. Effect of quality, porosity and density on the compression properties of cork. Holz Roh Werkst 66(4):295–301. https://doi.org/10.1007/s00107-008-0248-2

[2] Cork Quality Council. 2015. Industry Statistics. https://www.corkqc.com/pages/industry-statistics

[3] Fortes MA, Rosa ME, Pereira H. 2004. A Cortiça. IST Press, Lisboa

[4] Ghalem Amina, Barbosa Inés, Bluhraoua Rachid Tarik, Costa Augusta. 2018. Climate signal in cork-ring chronologies: case studies in southwestern Portugal and northwestern Algeria. Tree-ring research 74 (1): 15-27. https://doi.org/10.3959/1536-1098-74.1.15

[5] González-Hernández F, Gonzalez-Adrados JR, Garcia de Ceca JL .2012. Patente de Invención P200901750: Equipo para la medida de la fuerza de relajación de tapones tras el encorchado [Patent of Invention P200901750: Device for the measurement of the relaxation force of stoppers after corking].

Materials Research Forum LLC
https://doi.org/10.21741/9781644900413-4

[6] González Hernández, F., González Adrados, J.R., García de Ceca, J.L., Sánchez González, M. 2014. Quality grading of cork stoppers based on porosity, density and elasticity. Eur. J. Wood Prod. 72:149-156. https://doi.org/10.1007/s00107-013-0760-x

[7] Giunchi A, Versari A, Parpinello GP, Galassi S. 2008. Analysis of mechanical properties of cork stoppers and synthetic closures used for wine bottling. J Food Eng 88(4):576–580. https://doi.org/10.1016/j.jfoodeng.2008.03.004

[8] MAPAMA 1940 – 1971. ESTADISTICA FORESTAL DE ESPAÑA

[9] http://www.mapama.gob.es/es/desarrollo-rural/estadisticas/forestal_estadistica_forestal_1940_1971.aspx

[10] MAPAMA 2005_2013. ESTADISTICA FORESTAL DE ESPAÑA
http://www.mapama.gob.es/es/desarrollo-rural/estadisticas/forestal_anual_otros_aprovechamientos.aspx

[11] NORMA UNE-EN 12726 - 2000: Envases y embalajes. Boca para tapón de corcho con un diámetro de entrada de 18.5 mm para corchos y cápsulas de seguridad.

[12] Ovando Paola. Campos Pablo. Montero Gregorio. 2007. Forestaciones con encina y alcornoque en el área de la dehesa en el marco del Reglamento (CEE) 2080/92 (1993-2000). Revista Española de Estudios Agrosociales y Pesqueros nº 214 pp 173-186

[13] Pereira H. 2007. Cork: biology, production and uses. Elsevier, Oxford. https://doi.org/10.1016/B978-044452967-1/50013-3

[14] Prades C; Gómez-Sánchez, I; García-Olmo J; González-Hernández F;, González- Adrados JR. 2014. Application of VIS/NIR spectroscopy for estimating chemical, physical and mechanical properties of cork stoppers. Wood Sci Technol 48 (4): 811- 830. https://doi.org/10.1007/s00226-014-0642-3

[15] Rodney Keenan J. 2015. Climate change impacts and adaptation in forest management: a review. Annals of Forest Science 72:145–167. https://doi.org/10.1007/s13595-014-0446-5

Cork Science and its Applications II
Materials Research Proceedings 14 (2019) 28-34

Materials Research Forum LLC
https://doi.org/10.21741/9781644900413-5

Production of Cork Hollow Pieces by an Innovate Process Based on Rotational Moulding

Miguel Pestana*, Manuela Mendes[1], Luís Miranda[2] and António Corei Diogo[3]

*Instituto Nacional de Investigação Agrária e Veterinária, I.P. (INIAV, I.P.), Unidade de Tecnologia e Inovação, Av. da Republica, Quinta do Marquês, 2780-159 Oeiras, Portugal

†Robcork, Valorização de Produtos de Cortiça, SA, Rua Johnson Controls, 7300-062 Portalegre, Portugal

††Rotomoldagem, SA, Zona Industrial Vale da Goita, Paul, 2560-232 Torres Vedras, Portugal.

†††Instituto Superior Técnico (IST) – Universidade de Lisboa, Dep. Engenharia Química, Av. Rovisco Pais, 1049-001 Lisboa, Portugal

*miguel.pestan@inaiv.pt

Keywords: Recycling, Rotational Moulding, Composite Materials, Cork, Polyethylene

Abstract. A new technique of producing hollow cork composite parts is presented. It is mainly based on rotational moulding of cork powder and thermoplastic materials. Cork powder is a by-product from cork industry which is potentially dangerous (risk of fire and explosion); the incorporation of cork powder into the new products constitutes both an application of increased added value and a safe utilisation of cork powders. A series of mechanical tests on samples extracted from rotational moulded parts of the new cork composite are presented. Compression tests, creep and creep recovery tests are reported. There is evidence of toughening and softening (increased creep compliance) as the cork content increases; in both cases, the macrostructure of the composite contributes to the overall changes.

Introduction

Portugal is the biggest cork producer and exporter in the world (10^5 tonnes in 2010). It is estimated that cork waste (cork pellets and cork dust) from cork manufacture is about 20% to 30% of the raw material input [1]. The valorisation of such an amount is a big challenge and concern for cork industry. Cork agglomerates based on cork pellets found an increasing number of applications [2]. A number of prospective applications of cork pellets in different cork composites were also considered [3].

Valorisation of cork dust and cork pellets stands on looking for new applications with increased incorporation of added value. One example is project RotoCork (2011-15), which is a partnership among Robcork - Valorização de Produtos de Cortiça SA, Rotomoldagem SA, Instituto Nacional de Investigação Agrária e Veterinária, I.P. (INIAV, IP) and Instituto Superior Técnico - University of Lisbon (IST-UL), partially supported by Agencia de Inovação under COMPETE Program.

The main goal of project RotoCork is the production of hollow pieces in cork composites by non-standard variants of rotational moulding. Up to now, the standard way to produce hollow objects made of cork composites (e.g. cork agglomerate) was by excavating a massive solid block, previously obtained. This involves a protracted process which generates a huge amount of residues, which may attain 90% of the original weight. Manufacture of hollow parts in cork composites by non-standard variants of rotational moulding, with incorporation of cork dust and cork pellets, reduces the amount of secondary waste to some residual value which may be taken as negligible. Therefore, a direct outcome of RotoCork project is the upgrade of cork powder, a residue from cork industry, by its conversion to raw material for new products. Also, important savings in time and labour costs were accomplished. Two more accomplishments of the project must also be emphasized: first, a number of geometrical shape restrictions for hollow parts were removed, and second, the range of application of rotational moulding was extended to a new class of composite materials.

A number of tests were performed in order to get a detailed characterisation of the new products: physico-chemical tests, mechanical tests and so on.

The aim of this paper is to present the results of a number of mechanical tests involving creep under compression and creep recovery after compression. As a matter of fact, the performance of cork and cork composites is related to the way parts withstand compression stresses in different ways. A numerical simulation of the behaviour of cork agglomerates in compression and traction can be found in reference [4].

Creep tests were also considered. Creep tests very often cross the linear viscoelastic threshold. A discussion of non-linear creep effects e.g. in polyethylene (UHMWPE) can be found in reference [5].

Materials and methods

Several cork/polyethylene composite mouldings were produced by rotational moulding in different geometrical shapes, different formulations and different processing conditions. Tests were performed on rectangular plates extracted from the different parts. No shell testing will be reported here.

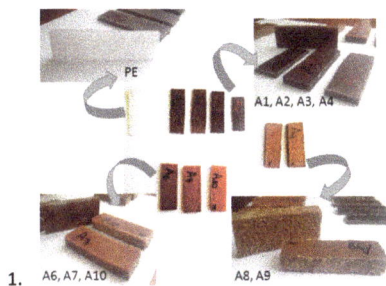

Figure 1 *– Images of the rectangular plates cut from different mouldings.*

For every moulding, at least three specimens were extracted. Some of them are shown in figure 1. In our nomenclature, PE refers to polyethylene, took as reference; A1, A2, A3 and A4 refer to PE + cork powder composites; A6, A7 and A10 refer to layered composites processed by sequential rotational moulding; A8 and A9 refer to PE and cork powder composites with high cork content. They were extracted from mouldings with different shapes, different compositions, and produced in different processing conditions.

Compression testing, and creep and creep recovery testing were performed in an Amsler Otto Wolpert-Werke GMBH D-6700 press. Typical parameters used in compression tests were: initial force (−180 N), compression speed (−120 N/s), sampling time (2 s). The maximum force was either (−40 kN) or (−65 kN). Creep/creep recovery tests, in compression, were performed by application of a trapezoidal force cycle to the specimens. In a trapezoidal compression cycle, a compression

Cork Science and its Applications II   Materials Research Forum LLC
Materials Research Proceedings **14** (2019) 28-34   https://doi.org/10.21741/9781644900413-5

ramp force is applied at constant rate, up to a maximum chosen value (e.g. −40 kN). The maximum compression force is kept for a given time (holding time ~5 min) and then removed. After removal of the compression force the recovery of the initial shape is monitored. After recovery, a new cycle may be started.

Results anda discussion
In this section, the experimental results obtained in compression (uniaxial compression) and creep/creep recovery tests will be presented and discussed.

Compression
Stress-strain curves were computed from the force-displacement data collected in the experiments. For the strain ranges considered here, most of the strain values are well beyond the infinitesimal strain limit, so the Hencky strain $\{\ln[l(t)/l(0)]\}$ was adopted as the strain measure. We recall that one of the main advantages of the use of Hencky strain measure is additivity.

Figure 2 displays the stress-strain curves for the specimens shown in figure 1. It is worth noting that the different stress-strain curves span a wide region of the stress-strain plane.

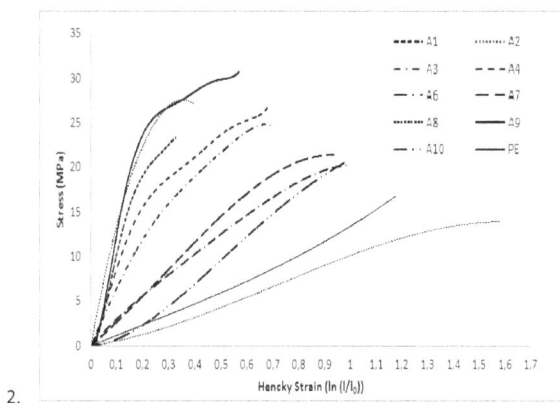

Figure 2 – *Stress-strain curves of the cork-PE composites and of PE, in compression.*

In figure 2 three different classes of behaviour emerge. The first one is constituted by low cork content HDPE/cork composites (A1, A2, A3 and A4), a second group is constituted by layered composites (A6, A7 and A10), and the third group includes samples A8 and A9 which are the ones with highest cork content. Average values of compression strength and compression modulus (Young modulus) for each class are presented in Table 1.

Cork Science and its Applications II
Materials Research Proceedings **14** (2019) 28-34

Materials Research Forum LLC
https://doi.org/10.21741/9781644900413-5

Table 1 – Average values of compressive strength and Young modulus of the different classes of cork/HDPE composites.

Samples	A1,A2, A3, A4	A6,A7, A10	A8, A9	PE
Compression Strength /MPa	26.3±2.1	21.1±0.9	15.1±1.9	30.8
Young Modulus/ MPa	94.6±29.5	25.8±4.0	15.2±1.6	139.3

Other important quantities are the changes of specimen thickness immediately after release of the compressive force, and after different times of recovery. The time of recovery is, by definition, the time lapse after compression release.

Table 2 – Deformation and recovery after one cycle compression test.

Samples	A1, A2, A3, A4	A6, A7, A10	A8, A9	PE
After test Hencky strain	-0.417	-0.885	-1.234	-0.572
Hencky strain after 50 min recovery	-0.120	-0.315	-0.399	-0.110
Recovery (Hencky)	0.297	0.570	0.835	0.462

There is a general pattern of increasing deformability (or increasing creep compliance) as the cork content increases. Also the recovery (after compression) shows a similar pattern.

Creep

Cork and thermoplastics are viscoelastic. Therefore, under a stress history $\sigma(t)$ which started t some finite time before the current time t, the time evolution of strain $\gamma(t)$ is

$$\gamma(t) = \int_{-\infty}^{t} dt'.J(t-t').\frac{d\sigma(t')}{dt'} = J_g.\sigma(t) - \int_{-\infty}^{t} dt'.\sigma(t').\frac{dJ_d(t-t')}{dt'}$$

(1)

where J(t) is a material function, the creep compliance. The creep compliance is the sum of the instantaneous compliance J_g and the delayed compliance $J_d(t)$:

$$J(t) = J_g + J_d(t)$$

(2)

For a linear viscoelastic material with a single retardation time, λ,

31

$$J_d(t) = J_d \cdot \left(1 - e^{-t/\lambda}\right) \tag{3}$$

An example of a trapezoidal stress wave is shown in figure 3.

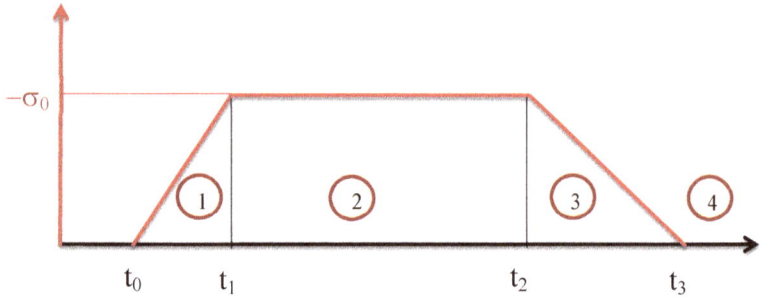

Figure 3 – *Trapezoidal stress wave (compression).*

The response of a linear viscoelastic material with a single retardation time, for which equation (3) holds, is given by equations (4-7). There is a direct correspondence between times $t_0 \ldots t_3$ in figure 1 and times $t_0 \ldots t_3$ in equations (4-7).

$$\frac{1}{\sigma_0} \cdot \gamma(t_0 < t < t_1) = \left(J_g + J_d\right) \cdot \frac{(t - t_0)}{(t_1 - t_0)} - J_d \cdot \lambda \cdot e^{-t/\lambda} \cdot \frac{\left(e^{t/\lambda} - e^{t_0/\lambda}\right)}{(t_1 - t_0)} \tag{4}$$

$$\frac{1}{\sigma_0} \cdot \gamma(t_1 < t < t_2) = \left(J_g + J_d\right) - J_d \cdot \lambda \cdot e^{-t/\lambda} \cdot \frac{\left(e^{t_1/\lambda} - e^{t_0/\lambda}\right)}{(t_1 - t_0)} \tag{5}$$

$$\frac{1}{\sigma_0} \cdot \gamma(t_2 < t < t_3) = -\left(J_g + J_d\right) \cdot \frac{(t - t_3)}{(t_3 - t_2)} - J_d \cdot \lambda \cdot e^{-t/\lambda} \cdot \left[\frac{\left(e^{t_1/\lambda} - e^{t_0/\lambda}\right)}{(t_1 - t_0)} - \frac{\left(e^{t/\lambda} - e^{t_2/\lambda}\right)}{(t_3 - t_2)}\right] \tag{6}$$

$$\frac{1}{\sigma_0} \cdot \gamma(t > t_3) = -J_d \cdot \lambda \cdot e^{-t/\lambda} \cdot \left[\frac{\left(e^{t_1/\lambda} - e^{t_0/\lambda}\right)}{(t_1 - t_0)} - \frac{\left(e^{t_3/\lambda} - e^{t_2/\lambda}\right)}{(t_3 - t_2)}\right] \tag{7}$$

The generalisation of equations (4-7) to a discrete spectrum of retardation times is a bit cumbersome but can be done without major difficulty; it will not be presented here for the sake of space. This generalisation of equations (4-7) for a linear viscoelastic medium with a discrete spectrum of retardation times was used in the computation of the creep compliance.

Figure 4 displays a typical response to a trapezoidal stress wave. Figure 5 displays the response to a sequence of trapezoidal stress waves.

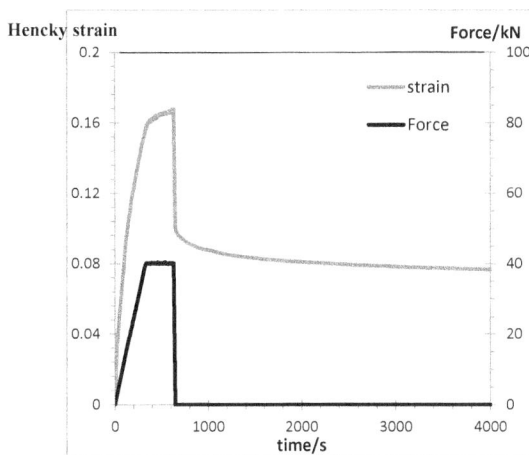

Figure 4 – *Response of one A7 specimen to a sequence of trapezoidal stress waves.*

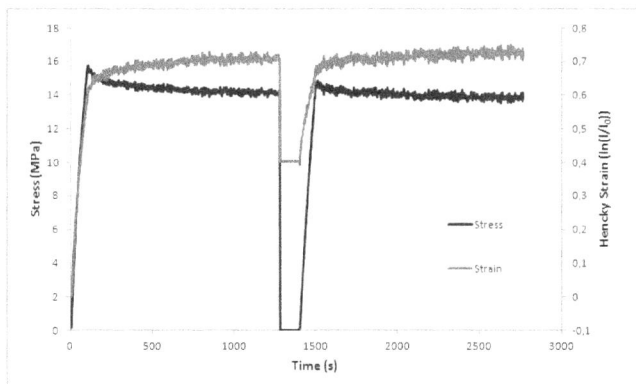

Figure 5 – *Response of one A7 specimen to a sequence of trapezoidal stress waves.*

In both cases a single retardation time is not enough to fit the time dependent creep compliance. Besides a short time process (of the order of a few seconds) which may be associated to wall bending, there is a long time process of the order of magnitude of an hour. The collapse of the cell walls introduces non-linearity and some irreversibility in the viscoelastic response. A more detailed analysis will be presented elsewhere.

Conclusions

Composites obtained from thermoplastics and cork powder may be processed by rotational moulding and they can provide hollow parts with minimum residues or by-products. This represents a substantial improvement when compared to the traditional way of excavating a massive solid block previously produced. Through the use of a number of improvements of basic rotational moulding technology, layered composites can also be produced.

According to the composition and processing variables, it is possible to span a wide region of the stress-strain diagram for compression. As a matter of fact, the introduction of cork softens the thermoplastic: in more technical and precise words, cork incorporation increases the creep compliance of the composite. Modulation of the changes in creep compliance can be achieved by simultaneous changes in the cork content and in the processing procedures.

Creep recovery depends on the amplitude of the compression force. At low amplitude values, linear behaviour is found. Linear (or quasi-linear) behaviour is recovered at high amplitudes, as a consequence of cork cells crushing; the retardation spectrum is nevertheless changed. At middle range amplitudes non-linear behaviour is found, much probably driven by cell walls buckling.

Acknowledgements

Project RotoCork is funded by Agência de Inovação under contract QREN number 21542, in the framework of COMPETE Program and Sistema de Incentivos à Investigação e Desenvolvimento Tecnológico do QREN: financial support is gratefully acknowledged.

References

[1] Cordeiro, N.; Belgacem, M.N.; Silvestre, A.J.D.; Neto, C.P.; Gandini, A. (1998). Cork suberin as a new source of chemicals. 1. Isolation and chemical characterization of its composition. International Journal of Biological Macromolecules 22, 71-80. https://doi.org/10.1016/S0141-8130(97)00090-1

[2] Pereira, H (2007). Cork: Biology, Production and Uses, Elsevier, Amsterdam. https://doi.org/10.1016/B978-044452967-1/50013-3

[3] Gil, L. (2009). Cork Composites: A Review. Materials, 2, 776-789. https://doi.org/10.3390/ma2030776

[4] Todo Bom, L.F.R. (2010). Comportamento à compressão e tracção da cortiça: estudo numérico. Dissertação de Mestrado em Engenharia Mecânica, Universidade de Aveiro.

[5] André, J.R.S. e Cruz-Pinto, J.J.C. (2005). Previsão do comportamento à fluência do polietileno de massa molecular ultra-elevada. Revista Iberoamericana de Polímeros, 6, 181-198.

Cork Science and its Applications II
Materials Research Proceedings **14** (2019) 35-40

Materials Research Forum LLC
https://doi.org/10.21741/9781644900413-6

A New 3D Print Technology Filament 100% Based in Natural Biological Sources and Cork Waste

VIEIRA, Flávia A.[a,1,2*], DA SILVA, Sara P. M [b,1,2] and DE OLIVEIRA, José M. [c,1,2]

[1]EMarT Group –Emerging Materials Research and Technologies - School of Design, Management and Production Technologies Northern Aveiro- ESAN, University of Aveiro, Estrada do Cercal, 449, 3720-509 Oliveira de Azeméis, Aveiro, Portugal

[2] Aveiro Institute of Materials- CICECO. University of Aveiro. Campus de Santiago,3810-193, Aveiro, Portugal

[a]flavia.vieira@ua.pt; [b]sarapms@ua.pt, [c] martinho@ua.pt

Keywords: Biobased and Biodegradable Cork-Composite, Cork Powder, Thermoplastic Starch Blends

Abstract. Lately, there is an increasing demand for natural and sustainable materials, which contributes to the natural appealing of cork uses. Cork industry produces around 30 % wt of residues in powder form. The main goal of this work is to use cork as reinforcing phase in a maize starch thermoplastic (TPS) matrix producing a cork-polymer composite (CPC) from 100 % natural sources. The starch thermoplastic matrices were produced using three different weight proportions of starch. Cork powders with granulometries ranging from 63µm-80µm were considered. The moisture absorption was evaluated during 16 weeks in a shelf-life storage experiment. The morphological characteristics were performed by SEM images and chemical profile by spectrometric and thermal analysis. Both TPS and CPC formulations were characterized in terms of moisture absorption. All infrared spectra were pre-treated and analyzed under multivariate statistics. The samples with lower moisture absorption and best group of results for chemical and morphological characterization were selected for further mechanical tests and 3D filaments production.

Introduction

In the last decade, the development of biobased and biodegradable plastic materials has been increasing, which tends to reduce its impact in the society and in the environment[1]. Starch, an agro-sourced polymer, can be used as a raw material and can contribute to produce bioplastics from natural and biological sources, mainly due to its wide availability, low cost and total recyclability without toxic waste [1]–[3]. Thermoplastic starch polymer (TPS) can be achieved by adding a plasticizer, such as glycerol or a swelling agent as water [4], [5], where the proportion of plasticizer and swelling agent can be, in general, 10%-40% wt % based on total amounts (in mass) of components. When heated in water, starch granules swell. The starch undergo a transition process, in which their constituents (amylose and amylopectin) leach out until the total disruption of the granules and the formation of a polymers-in-solution mixture. So, the starch began the gelatinization process and because of these destructuration it can behave as a thermoplastic polymer. The crystallinity of starch granules disappears during this process and an amorphous material is obtained. Under these conditions it can be processed by conventional techniques [6]–[8]. Starch moisture content changes according to the variation in the relative humidity of the atmosphere in which it is stored. To overcome these limitations, TPS formulations can be mixed with other natural [9] or synthetic polymers[10] and improve its own drawback [11]. Another way to surpass TPS moisture problems and its poor mechanical behavior is by reinforcing the blends with other natural raw material, as cork. Cork has a unique alveolar structure, low density and good mechanical properties, good insulation properties and allows high levels of filling. They are readily available materials, recyclable and non-toxic[12]. Some of these qualities distinguish cork from other wood materials [13], [14]. The incorporation of cork powder in TPS blends can be an effective approach to develop new sustainable materials.

Materials and Methods

Materials

Three formulations of TPS blends with starch, glycerol and water were performed. Starch (Maizena®) was purchased in local market; the glycerol was obtained from Himedia with 99% of purity and distillated water were used. The addition of cork powder, supplied by a Portuguese cork producer, remained equal to 11% (wt %) for all 3 CPC formulations. Cork powder was fractionated through sieving (Retsch, Germany) and the average particle size was determined. In this work, cork powder granulometry ranged between 63-80 μm was used. The formulations were named in accordance with the higher glycerol amount: TPS-1, formulation with 40% of glycerol in its constitution; TPS-2, formulation with 30% of glycerol; and, TPS-3, formulation with lower amounts of glycerol (20%). Consequently, the CPC-1, CPC-2 and CPC-3 are the cork-polymer composites with 40%, 30% and 20% of glycerol, respectively. All blends compositions are shown in Table 1.

Table 1- Compositions of TPS and CPC formulations.

Formulations	Percentage composition (wt/wt)%			
	Starch	Glycerol	Water	Cork
TPS-1	50	40	10	-
TPS-2	50	30	20	-
TPS-3	50	20	30	-
CPC-1	44	36	9	11
CPC-2	44	26	18	11
CPC-3	44	18	27	11

Methods

In a Brabender type machine, all formulations (Table 1) were prepared by mixing the raw materials during 10 minutes at 130 °C and 100 rpm. All samples were transformed into pellets using a granulator. A visual appearance observations and touch behavior were done before the chemical and morphological characterization process. The moisture content (wt%) was measured in a moisture analyzer equipment (Radwag) until weight constantly at temperature 120 °C. The moisture absorption ability was determined after 16 weeks of storage at non-controlled environment. Morphological characterization was performed using a SEM equipment (Hitachi SU-70) – the samples were covered with palladium gold and images were taken with four magnifications (40X; 100X; 300X and 1000X). Chemical characterization was performed using a Bruker 27 with golden gate (ATR), where the infrared spectra were acquired with 256 scans ,4 cm^{-1} resolution and a wavelength ranging from 3500 to 400 cm^{-1}. The measured spectra were normalized, baseline corrected, and transferred via a JCAMP.DX format into the data analysis. A free-software for Principal Components Analysis (PCA) (PAST program) was used. Previously, to PCA analysis, each spectrum within the 3000-600 cm-1 region was standard normal deviates corrected[15]. The

Cork Science and its Applications II
Materials Research Proceedings 14 (2019) 35-40

Materials Research Forum LLC
https://doi.org/10.21741/9781644900413-6

differential scanning calorimetry (DSC) was used to measure the thermal transitions of TPS CPC samples. The test was performed with a Shimadzu-50 differential scanning calorimeter equipment, fitted with a nitrogen based cooling system. All the measurements were performed in the temperature range of 25 to +200 °C at a heating rate of 10 °C/min.

Results and Discussion

Just by looking and handing TPS samples, it was possible to observe that when the presence of glycerol decreased, samples gloss and adhesiveness decreased in the same way. Probably, this was caused by the overplasticization effect promoted by glycerol. In **Figure 1**, this peculiar aspect can be confirmed with the appearance of a granular surface from pellets. The overplasticization is the cause of the presence of porous structures and starch granules below the thin plasticized layer in TPS-1. An inefficient temperature and shearing could be the cause of these defects for this particular TPS formulation. For TPS-2 and TPS-3, the plasticization was complete.

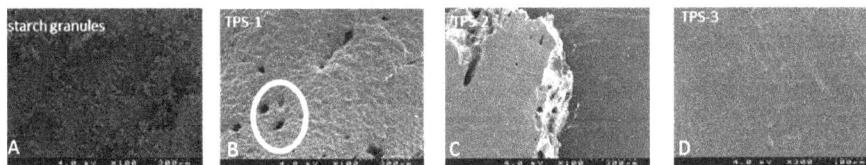

Figure 1. SEM images with 100X ampliations for: starch granules (A); TPS-1 (B); TPS-2 (C) and TPS-3 (D) with 300 X ampliation.

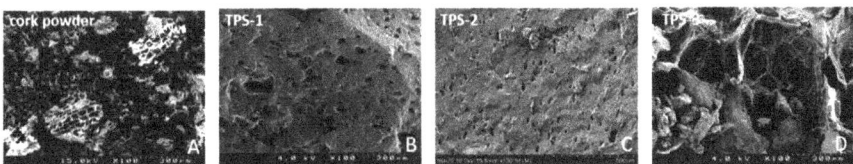

Figure 2. SEM images with 100X ampliations for: cork powder (A); CPC-1(B); CPC-2 (C) and CPC-3 (D).

The addition of cork powder (**Figure 2**) promoted the presence of porous structures. However, in CPC-3, the alveolar cork structure below the thin starch thermoplastic layer was more visible than CPC-1 and CPC-2. In fact, it seems that the production of this composite was not complete.

The FTIR spectra of TPS blends and CPC featured absorption bands corresponding to the functional groups of starch and glycerol. Bands at 920 cm^{-1}, 1022 cm^{-1} (their positions did not seem to change in all spectra observed), 1148 cm^{-1} (CO stretching in the anhydroglucose ring of starch), 1648 cm^{-1} (bound water), 3277 cm^{-1} (OH groups), 2914 cm^{-1} (CH stretching) and 1423 cm^{-1} (glycerol) [11], [16]. The broad band at 3370–3355cm^{-1} was attributed to the vibrational stretches of hydroxyl groups in starch. This band shifted to lower wavenumber with the addition of glycerol, indicating the formation of hydrogen bonds between starch and glycerol. The increase of this band intensity became more obvious as glycerol content increased, due to the formation of new hydrogen bonds [17]–[19].

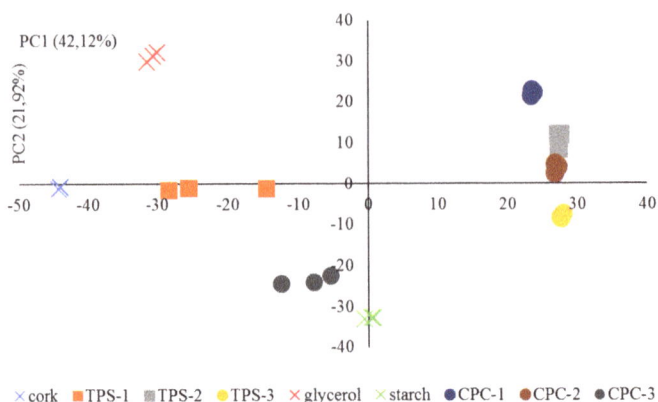

Figure 3. *Figure 3 represents the PCA scores scatter plot of the FTIR dataset of all samples (TPS and CPC and their raw-materials).*

Figure 3 shows a PCA scores scatter plot using FTIR spectra dataset of all samples (TPS-1; TPS-2; TPS-3; CPC-1; CPC-2 and CPC-3) and their raw-materials components (starch, glycerol and cork). The classification of the samples resulting from the PCA analysis of the infrared dataset revealed that PC1 vs. PC2 contains 64% of the dataset variability. Besides, a clear segregation of the raw-materials, TPS-1 and CPC-3 is noticed as two groups were detected along PC1 axis. The loadings seem to indicate that the TPS fingerprint peaks can be associated to differences between them. The loadings identify the TPS formation in PC1 positive axis. The results show that multivariate analysis assist to identify differences in behavior of samples during the extrusion process and the TPS/CPC production, conducting to a better characterization of the differences among them. On the other hand, the DSC tests (not shown) indicate an improvement in thermal behavior for CPC-2. While TPS-2 has 77.5°C of melting point of fusion, with cork powder addition (CPC-2) the melting point increased for 150°C. This behavior was unexpected because for others formulations (TPS-1; TPS-3, CPC-1 and CPC-3) the thermal profile kept the same tendency. In all samples, the melting point was 77°C. The moisture absorption experiment was conducted to evaluate the moisture percentage after the extrusion production of samples and in pellets storage.

The determination of moisture percentage of samples is important for predicting theirs stability during storage. The shelf-life of the CPC, in ambient conditions, is dependent on their moisture uptake. In **Figure 4A**, with the decrease of glycerol content, the moisture absorption decreased too. With the addition of cork powder, the opposite behavior were seen. Cork powder appears to minimize the glycerol tendency for hygroscopy.

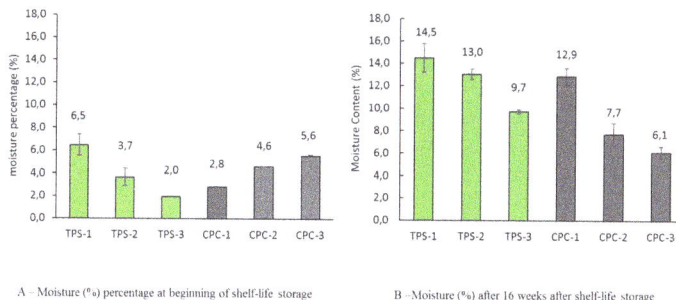

A – Moisture (%) percentage at beginning of shelf-life storage   B – Moisture (%) after 16 weeks after shelf-life storage

Figure 4: *4A represents the graphic of moisture absorption after the sample's preparation and 4B represents the moisture absorption after 16 weeks of storage.*

Conclusions

A biodegradable thermoplastic starch was prepared from glycerol, water and cornstarch. The same success was observed with CPC development using cork powder as reinforcing phase for TPS. The moisture absorption ability of TPS decreased with the addition of cork powder residues, which improved their chemical behavior and, consequently, their storage life capacity. In general, the TPS-2 and CPC-2 plasticized with 30% of glycerol displayed the best overall mixtures. Nevertheless, improvements in these formulations are needed, in order to consider them as candidate for 3DP filaments production.

References

[1]   Zia-ud-Din, H. Xiong, and P. Fei, Physical and chemical modification of starches: A review, *Crit. Rev. Food Sci. Nutr.*, 57, 12, (2017), 2691–2705. https://doi.org/10.1080/10408398.2015.1087379

[2]   J. Waterschoot, S. V. Gomand, E. Fierens, and J. A. Delcour, Starch blends and their physicochemical properties, *Starch - Stärke*, 67, 1–2, (2015), 1–13. https://doi.org/10.1002/star.201300214

[3]   T. Mekonnen, P. Mussone, H. Khalil, and D. Bressler, Progress in bio-based plastics and plasticizing modifications, *J. Mater. Chem. A*, 1, (2013), 13379. https://doi.org/10.1039/c3ta12555f

[4]   A. A. Aydın and V. Ilberg, Effect of different polyol-based plasticizers on thermal properties of polyvinyl alcohol:starch blends, *Carbohydr. Polym.*, 136, (2016), 441–448. https://doi.org/10.1016/j.carbpol.2015.08.093

[5]   M. G. A. Vieira, M. A. da Silva, L. O. dos Santos, and M. M. Beppu, Natural-based plasticizers and biopolymer films: A review, *Eur. Polym. J.*, 47, 3, (2011), 254–263. https://doi.org/10.1016/j.eurpolymj.2010.12.011

[6]   M. R. Area *et al.*, Corn starch plasticized with isosorbide and filled with microcrystalline cellulose: Processing and characterization, *Carbohydr. Polym.*, 206, (2019), 726–733. https://doi.org/10.1016/j.carbpol.2018.11.055

[7]   R. F. Tester, J. Karkalas, and X. Qi, Starch—composition, fine structure and architecture, *J. Cereal Sci.*, 39, 2, (2004), 151–165. https://doi.org/10.1016/j.jcs.2003.12.001

[8]   Y. Zhang, C. Rempel, and Q. Liu, Thermoplastic Starch Processing and Characteristics—A

Review, *Crit. Rev. Food Sci. Nutr.*, 54, 10, (2014), 1353–1370.
https://doi.org/10.1080/10408398.2011.636156

[9] K. M. Dang and R. Yoksan, Development of thermoplastic starch blown film by incorporating plasticized chitosan, *Carbohydr. Polym.*, 115, (2015)575–581. https://doi.org/10.1016/j.carbpol.2014.09.005

[10] L. Tan, Q. Su, S. Zhang, and H. Huang, Preparing thermoplastic polyurethane/thermoplastic starch with high mechanical and biodegradable properties, *RSC Adv.*, 5, 98, (2015), 80884–80892. https://doi.org/10.1039/C5RA09713D

[11] J. F. Mendes *et al.*, Biodegradable polymer blends based on corn starch and thermoplastic chitosan processed by extrusion, *Carbohydr. Polym.*, 137, (2016), 452–458. https://doi.org/10.1016/j.carbpol.2015.10.093

[12] S. P. Silva, M. A. Sabino, E. M. Fernandes, V. M. Correlo, L. F. Boesel, and R. L. Reis, Cork: properties, capabilities and applications, 2005. https://doi.org/10.1179/174328005X41168

[13] I. M. Aroso, A. R. Araújo, R. A. Pires, and R. L. Reis, Cork: Current Technological Developments and Future Perspectives for this Natural, Renewable, and Sustainable Material, *ACS Sustain. Chem. Eng.*, 5,. 12, (2017), 11130–11146. https://doi.org/10.1021/acssuschemeng.7b00751

[14] S. P. Magalhães da Silva, P. S. Lima, and J. M. Oliveira, Rheological behaviour of cork-polymer composites for injection moulding, *Compos. Part B Eng.*, 90, (2016) 172–178. https://doi.org/10.1016/j.compositesb.2015.12.015

[15] J. Čopíková *et al.*, Influence of hydration of food additive polysaccharides on FT-IR spectra distinction, *Carbohydr. Polym.*, 63, 3, (2006), 355–359. https://doi.org/10.1016/j.carbpol.2005.08.049

[16] Y. Zhong and Y. Li, Effects of glycerol and storage relative humidity on the properties of kudzu starch-based edible films, *Starch - Stärke*, 66, 5–6, (2014), 524–532. https://doi.org/10.1002/star.201300202

[17] H.-Y. Wang and M.-F. Huang, Preparation, characterization and performances of biodegradable thermoplastic starch, *Polym. Adv. Technol.*, 18, 11, (2007). https://doi.org/10.1002/pat.930

[18] H. A. Pushpadass, D. B. Marx, and M. A. Hanna, Effects of Extrusion Temperature and Plasticizers on the Physical and Functional Properties of Starch Films, *Starch - Stärke*, 60, 10, (2008), 527–538. https://doi.org/10.1002/star.200800713

[19] R. Shi *et al.*, Ageing of soft thermoplastic starch with high glycerol content,*J. Appl. Polym. Sci.*, 103, 1, (2007) 574–586. https://doi.org/10.1002/app.25193

Cork Science and its Applications II
Materials Research Proceedings **14** (2019) 41-46

Materials Research Forum LLC
https://doi.org/10.21741/9781644900413-7

Crashworthiness of Agglomerated Cork Under the Influence of Extremely Low and High Temperatures

Johannes Wilhelm[1, a], Pawel Kaczynski[1, b] Mariusz Ptak[1, c *], Fabio A.O. Fernandes[2, d] and Ricardo J. Alves de Sousa[2, e]

[1] Wroclaw University of Science and Technology, Faculty of Mechanical Engineering, Łukasiewicza 7/9, 50-371 Wrocław, Poland

[2] Center for Mechanical Technology and Automation, Universidade de Aveiro, Campus Santiago, 3810-193, Aveiro, Portugal

[a]johannes.wilhelm@pwr.edu.pl, [b]pawel.kaczynski@pwr.edu.pl, [c]mariusz.ptak@pwr.edu.pl, [d]fabiofernandes@ua.pt, [e]rsousa@ua.pt

Keywords: Agglomerated Cork, Natural Composites, Extreme High and Low Temperatures, High-Energy Impacts, Crashworthiness

Abstract. Cork material is utilized nowadays in a wide variety of applications due to its excellent shock absorption, thermal and acoustic insulation properties. Especially, this applies to agglomerated cork, which is acting in applications of even highly demanded dimensional stability nearly isotropic due to the random orientation of its grains and offers dominantly viscoelastic behaviour with almost zero Poisson's ratio. As the interest in the outdoor application of cork material increases, the assessment of its performance under extremely low and high temperatures is inevitable. The research addresses this topic for five different types of cork agglomerates and assesses their capability to withstand an impact energy of 500 J from sub-zero temperatures of -30°C up to 100°C. Thereby, the research covers a full span of working circumstances, including automotive and aeronautics and their passive safety applications. The results show dependent on the tested temperature significant variations in the amount of absorbed energy. Hence, the attention of product designers and developers is called to consider the temperature-dependent performance, when it comes to dimensioning of product for extreme weather conditions.

Introduction

Searching for natural materials, able to be property-tailored and substituting synthetic materials in engineering applications and products gained to be a top-priority task among researchers. Current governmental policies as well as environmental-friendly tendencies of manufacturers and consumers to choose eco-friendly, partially even recyclable alternatives to common market solutions are justifying the done effort in this field. Showing a related and promising potential in this concern, cork — firstly as agglomerate — has been integrated in solutions or technical designs due to its outstanding mechanical, thermal and acoustic properties. Thereby, the range of application is wide, as it can be found within passive safety devices as well as in general automotive sector, even in demanding applications as the aerospace ones [1], [2]. As the temperature spreads within these fields of application, special regard concerning the performance under extreme temperatures has to be taken.

Accordingly, the characterization of cork material has been center of interest for several studies within the last decades and spawned among others several models for the prediction of corks mechanical properties under compressive loading. Some studies were based thereby on data mining and machine learning techniques [3]. Apart from this, only a few recent studies and works concentrated on corks behavior under impact [4], [5], [6] and [7]. Even if these studies revealed micro-agglomerated corks capabilities to improve components regarding their crashworthiness and energy absorption under impact or blast loads, the temperature range was thereby only rarely covering extremely low or high values. Hence, the authors of this study performed initially research by assessing the influence of temperature to agglomerated cork material for impacts with energies in the range of 120 to 850 J for temperatures between 21 and 50°C [7]. As the chemical stability of cork showed differences for the evaluated range of 150 to 500°C [8], the presented study is intended to enrich the research of agglomerated corks mechanical behavior in the temperature range of -30 to 100°C, as especially the application in passive safety devices demands by its variety of possible environmental conditions over the globe similar temperature values.

Materials and Models
As given by the introduction, micro-agglomerated cork with a density around 216 kg/m^3 was proposed as a promising option for substituting EPS as shock absorbing layer in passive safety devices. Consequently, the authors of the presented work chose for their study several cork types with densities up to this value, also with differing grain sizes. An overview of all chosen cork types is given in Table 1.

Table 1: Cork material data and the range of tested temperatures

Cork type	Cork type/Density [kg/m^3]/ Grain size [mm]	Temperature [°C]
AC168	Micro-agglomerated /168/ 1-5	-30; -15, 0, 100
AC199A	Micro-agglomerated /199 /0.5-2	-30; -15, 0, 24, 100
AC199B	Micro-agglomerated / 199/ 0.5-1	-30; -15, 0, 100
AC216	Micro-agglomerated / 216 /2-4	-30; -15, 0, 24, 100
EC159	Expanded / 159 /4-10	-30; -15, 0, 24, 100

The tested specimens were prepared concerning their temperature by the use of a furnace equipped with a heater respectively connected nitrogen cooling and induced circulation, which enabled to keep the temperature constant over the tests. The specimen's temperature was supervised and verified using a K-type thermocouple placed in the core of a separate specimen.

All dynamic impact tests were performed using the instrumented uniaxial drop-weight tower Instron Dynatup 9250HV. For each temperature and each cork type, three samples with the cross-section of 50 mm x 50 mm and the height of 60 mm were placed central on a lower anvil and impacted by a flat tup (diameter: 50 mm) with an initial impact velocity of 9.2 m/s and a total mass of 11.2 kg. Out of this setup, the experiment is considered to be repeatable. Data was recorded by a load cell with a sample rate of 204.8 kHz and visually by the use of a high-speed camera Phantom V12 at a frame rate of 10000 fps. The analysis of the video recordings was performed within the dedicated software TEMA Motion, which enables to obtain displacement-time data for each of the tested specimens according to beforehand installed high-contrast markers on anvil and machine's impactor. The revealed data was correlated by using National Instruments LabVIEW software and adjusting the data sets regarding their differing sample rates.

Cork Science and its Applications II
Materials Research Proceedings **14** (2019) 41-46

Materials Research Forum LLC
https://doi.org/10.21741/9781644900413-7

Results and discussion

During the tests it was observed, that the samples undergo large uniaxial strain and reaching their maximum compressive displacement approximately 8 ms after the initial contact with the tup. The overall temperature range of -30 up to 100°C implied a huge difference in the behavior under compression, even if — by considering cork materials Poisson's ratio of 0 up to 0.15 [9] — the expected behavior of corks cell walls to start to collapse and buckle was observed. For the temperature of 100°C, AC199A and AC216 were observed to be less stiff comparing to the results under -30°C. Moreover, at 100°C in nearly every tested specimen cracks were observed and a "barrel"-effect appeared, where the sides' walls were partially or completely separated from the specimen (Fig. 1a), which did not occur at -30°C. It has to be pointed on the EC159 samples, which did not show this behavior due to their earlier destruction. Their grains are separated relatively easy, as this material contains no additional binder and consequently loses its integrity after losing the low adhesive capabilities during an impact scenario.

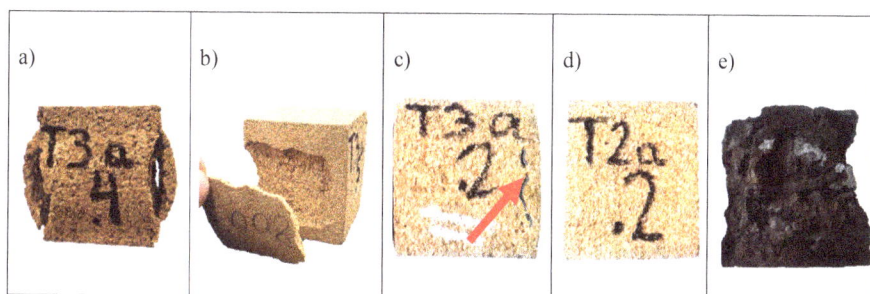

Fig. 1: Impacted cork specimen:
500 J at 100°C: a) AC168, b) AC199B, c) AC216 (crack marked by red arrow);
500 J at -30°C: d) AC216, e) EC159

In order to compare all the tested specimens, a deflection threshold of 38 mm was installed and used, as all specimen except EC were able to withstand up to this deflection. For EC the energy level was decreased — also according to previously gained experience — in order to check its crashworthiness in its lower energy-absorbing level compared to AC.

The results proved, that the major factor, which is affecting the crashworthiness of the used samples, is the temperature, hence the temperature of the application also. Fig. 2 is showing this influence for all testes cork types at the deflection of 38 mm.

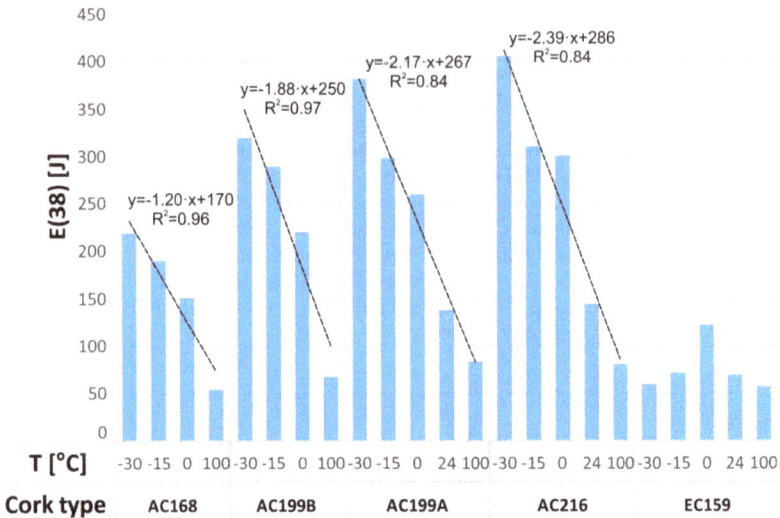

Fig. 2: The influence of temperature on the absorbed energy at 38 mm deflection after 500 J impact

Fig. 2 reveals, that increasing the temperature from -30°C to a value of 100°C was able to decrease the level of absorbed energy by more than 75 percent. The slope of this behavior is getting visible by the introduced trend lines in Fig. 2 and changes according to the density of the tested cork type. A higher density lead to a more intense slope. Oppositionally, EC is not showing this trend, as it is damaged upon the impact. Nonetheless, regarding the influence of temperature to the level of absorbed energy, cork layers exemplarily in passive safety devices hence need to be adjusted according to their envisaged environmental conditions of application. As Fig. 2 showed over the tested temperature range a linear influence, manufacturers can predict the energy absorbing capabilities and modify their products accordingly.

Although the temperature influence was observed for different cork material densities, the influence was differing for two cork materials which were specified by equal density but composed by grains with different size. The crashworthiness was observed to be significantly affected by the grain size. The phenomenon is related also to the binder — which was here polyurethane glue — the impact energy and the temperature during the impact test. Moreover, the amount of glue in the AC199B specimen was recognized as higher than in AC199A, as the grain size in AC199A was bigger. Furthermore, and due to its viscoelastic and viscoplastic behavior, the specimen showed up with thermal effects after being subjected to impact energies of 500 J. In this concern it needs to be mentioned, that cork material grains are very good heat insulators, characterized by a small heat capacity [10], if one is comparing them to the used polyurethane binder. Reference [7] stated, that the compression of air inside the cork specimen has to be seen as an adiabatic process. Conditioning the specimen to 100°C and keeping them on this temperature leads the glue to have a higher viscosity and decreasing adhesive capabilities. The cork grains as small heat capacitors then are causing the binder to be heated during the compression process. Consequently, the glues viscosity increases and the adhesive capabilities are decreased, ending in an increased grain separation at the microstructure level during the impact scenario. This damage of the structural integrity reduces in a consequence the energy absorption capabilities, for the AC199B compared to AC199A for deflection up to 60% of the specimens' height. The authors observed after this deflection of 35 mm, that the absorbed energy ratio for AC199A compared to AC199B became less than 1 [J/J]. The authors explain this, that AC199A's larger granules are showing tendencies to be more brittle and hence fracturing under high compression. Post-impact inspections of the specimen revealed thereby corresponding micro fractures in the large grains of AC199A and confirm the findings of reference [7].

Basing on the presented findings, the authors marked within the force-deflection curves of all tested AC199A the levels of absorbed energy for 100 J, 200 J and 300 J. It figured out, that the force level — visualised in Fig. 3 — is inversely proportional to the applied temperature.

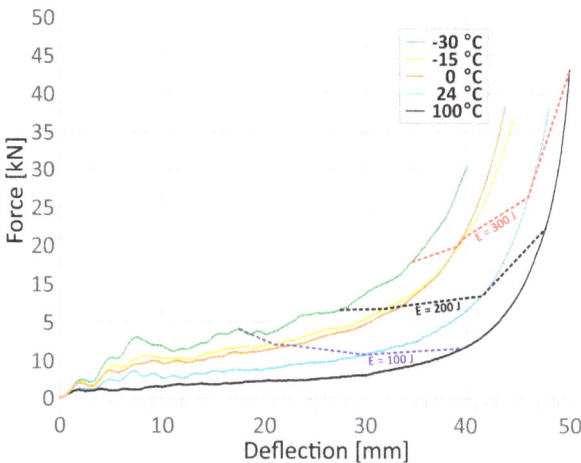

Fig. 3: Force-deflection curve of AC199A cork with the specific amount of absorbed energy

While the marked niveau of 300 J for the different temperatures is rising exponentially, the curve for 100 J of absorbed energy is over the temperature range rather flat. The result needs to be seen in an increased temperature influence on low-energy impacts. The authors stated initially, that for the technical design of passive safety devices — that are including an absorption liner made of a specific cork material — the energy level and the temperature in terms of the environmental conditions inevitably have to be considered. The presented finding for AC199A cork material assists furthermore in the choice of the mandatory layer thickness of this absorption liner.

Conclusions

In summary, the authors recapitulate all findings:

1. The energy absorption capabilities according to an impact is dependent on the cork materials density and the applied temperature. By increasing the specimens' density, the influence of the temperature to the absorbed energy rises as well.

2. The authors were able to confirm their previous gained results, that were presented in Ptak et al. (2017). Choosing a specific grain size drives also the amount of particle binder. It was visible, that a dropping reaction force concerns to the amount of used glue, which especially concerns the specimen, that were conditioned at 100°C. The value of the absorbed impact energy was by this directly influenced.

3. Cork specimens performance — especially for high-energy impacts — is significantly affected by the applied temperature.

References

[1]   J. M. Silva, T. C. Devezas, A. Silva, L. Gil, C. Nunes, and N. Franco, "Exploring the Use of Cork Based Composites for Aerospace Applications," *Mater. Sci. Forum*, vol. 636–637, pp. 260–265, Jan. 2010. https://doi.org/10.4028/www.scientific.net/MSF.636-637.260

[2]   D. Tchepel, F. A. O. Fernandes, O. Anjos, and R. Alves de Sousa, "Mechanical Properties of Natural Cellular Materials," in *Reference Module in Materials Science and Materials Engineering*, Elsevier, 2016. https://doi.org/10.1016/B978-0-12-803581-8.04056-X

[3]   Á. García, O. Anjos, C. Iglesias, H. Pereira, J. Martínez, and J. Taboada, "Prediction of mechanical strength of cork under compression using machine learning techniques," *Mater. Des.*, vol. 82, pp. 304–311, Oct. 2015. https://doi.org/10.1016/j.matdes.2015.03.038

[4]   C. P. Gameiro and J. Cirne, "Dynamic axial crushing of short to long circular aluminium tubes with agglomerate cork filler," *Int. J. Mech. Sci.*, vol. 49, no. 9, pp. 1029–1037, 2007. https://doi.org/10.1016/j.ijmecsci.2007.01.004

[5]   M. Paulino and F. Teixeira-Dias, "An energy absorption performance index for cellular materials – development of a side-impact cork padding," *Int. J. Crashworthiness*, vol. 16, no. 2, pp. 135–153, 2011. https://doi.org/10.1080/13588265.2010.536688

[6]   F. A. O. Fernandes, R. J. S. Pascoal, and R. J. Alves de Sousa, "Modelling impact response of agglomerated cork," *Mater. Des.*, vol. 58, pp. 499–507, 2014. https://doi.org/10.1016/j.matdes.2014.02.011

[7]   M. Ptak, P. Kaczynski, F. A. O. Fernandes, and R. J. A. de Sousa, "Assessing impact velocity and temperature effects on crashworthiness properties of cork material," *Int. J. Impact Eng.*, vol. 106, pp. 238–248, 2017. https://doi.org/10.1016/j.ijimpeng.2017.04.014

[8]   F. A. O. Fernandes, J. P. Tavares, R. J. Alves de Sousa, A. B. Pereira, and J. L. Esteves, "Manufacturing and testing composites based on natural materials," *Procedia Manuf.*, vol. 13, pp. 227–234, 2017. https://doi.org/10.1016/j.promfg.2017.09.055

[9]   R. T. Jardin, F. A. O. Fernandes, A. B. Pereira, and R. J. Alves de Sousa, "Static and dynamic mechanical response of different cork agglomerates," *Mater. Des.*, vol. 68, pp. 121–126, Mar. 2015. https://doi.org/10.1016/j.matdes.2014.12.016

[10] H. Pereira, "Chemical composition and variability of cork from Quercus suber L.," *Wood Sci. Technol.*, vol. 22, no. 3, pp. 211–218, 1988. https://doi.org/10.1007/BF00386015

Keyword Index

.

www.ingramcontent.com/pod-product-compliance
Lightning Source LLC
Chambersburg PA
CBHW071519210326
41597CB00018B/2819